Routing Congestion in VLSI Circuits

Estimation and Optimization

Series on Integrated Circuits and Systems

Series Editor: Anantha Chandrakasan
Massachusetts Institute of Technology
Cambridge, Massachusetts

Routing Congestion in VLSI Circuits: Estimation and Optimization
Prashant Saxena, Rupesh S. Shelar, Sachin Sapatnekar
ISBN 978-0-387-30037-5, 2007

Ultra-Low Power Wireless Technologies for Sensor Networks
Brian Otis and Jan Rabaey
ISBN 978-0-387-30930-9, 2007

Sub-Threshold Design for Ultra Low-Power Systems
Alice Wang, Benton H. Calhoun and Anantha Chandrakasan
ISBN 978-0-387-33515-5, 2006

High Performance Energy Efficient Microprocessor Design
Vojin Oklibdzija and Ram Krishnamurthy (Eds.)
ISBN 978-0-387-28594-8, 2006

Abstraction Refinement for Large Scale Model Checking
Chao Wang, Gary D. Hachtel, and Fabio Somenzi
ISBN 978-0-387-28594-2, 2006

A Practical Introduction to PSL
Cindy Eisner and Dana Fisman
ISBN 978-0-387-35313-5, 2006

Thermal and Power Management of Integrated Systems
Arman Vassighi and Manoj Sachdev
ISBN 978-0-387-25762-4, 2006

Leakage in Nanometer CMOS Technologies
Siva G. Narendra and Anantha Chandrakasan
ISBN 978-0-387-25737-2, 2005

Statistical Analysis and Optimization for VLSI: Timing and Power
Ashish Srivastava, Dennis Sylvester, and David Blaauw
ISBN 978-0-387-26049-9, 2005

Prashant Saxena
Rupesh S. Shelar
Sachin S. Sapatnekar

Routing Congestion in VLSI Circuits

Estimation and Optimization

Prashant Saxena
Synopsys, Inc.
Hillsboro, OR, USA

Rupesh S. Shelar
Intel Corporation
Hillsboro, OR, USA

Sachin S. Sapatnekar
University of Minnesota
Minneapolis, MN, USA

Routing Congestion in VLSI Circuits: Estimation and Optimization

ISBN 978-1-4419-4013-1 e-ISBN 978-0-387-48550-8
e-ISBN 0-387-48550-3

Printed on acid-free paper.

9 8 7 6 5 4 3 2 1

springer.com

To my parents Kailash and Savitri, and my wife Priti.
– Prashant

To my mother Keshar and siblings Sandhya and Tushar, and the memories of my father.
– Rupesh

To my family.
– Sachin

Preface

With dramatic increases in on-chip packing densities, routing congestion has become a major problem in integrated circuit design, impacting convergence, performance, and yield, and complicating the synthesis of critical interconnects. The problem is especially acute as interconnects are becoming the performance bottleneck in modern integrated circuits. Even with more than 30% of white space, some of the design blocks in modern microprocessor and ASIC designs cannot be routed successfully. Moreover, this problem is likely to worsen considerably in the coming years due to design size and technology scaling.

There is an inherent tradeoff between choosing a minimum delay path for interconnect nets, and the need to detour the routes to avoid "traffic jams"; congestion management involves intelligent allocation of the available interconnect resources, up-front planning of the wire routes for even distributions, and transformations that make the physical synthesis flow congestion-aware. The book explores this tradeoff that lies at the heart of all congestion management, in seeking to address the key question: how does one optimize the traditional design goals such as the delay or the area of a circuit, while still ensuring that the circuit remains routable? It begins by motivating the congestion problem, explaining why this problem is important and how it will trend. It then progresses with comprehensive discussions of the techniques available for estimating and optimizing congestion at various stages in the design flow.

This text is aimed at the graduate level student or engineer interested in understanding the root causes of routing congestion, the techniques available for alleviating its impact, and a critical analysis of the effectiveness of these techniques. The scope of the work includes metrics and optimization techniques for congestion at various stages of the VLSI design flow, including the architectural level, the logic synthesis and technology mapping level, the placement phase, and the routing step. This broad coverage is accompanied by a critical discussion of the pros and cons of the different ways in which one can minimize the ill-effects of congestion. At the same time, the book attempts to highlight further research directions in this area that appear promising.

Although this book is not meant to be an introductory text to VLSI CAD, we have tried to make it self-contained by providing brief primers that go over the classical techniques in routing, placement, technology mapping and logic synthesis, before diving into discussions on how these techniques may be modified to mitigate congestion. Our coverage focuses on congestion issues dealing primarily with standard cell based designs. In particular, the models and optimization methods that pertain specifically to field-programmable gate arrays (FPGAs) have not been explicitly addressed in this book.

Acknowledgments

It is said that no man is an island; similarly, no book is an island either. There are numerous people who have indirectly influenced this text. These include our mentors and colleagues over the years, as well as the many outstanding researchers who have left their mark on the field of VLSI CAD in general and physical synthesis in particular, and we owe a great debt to all of them.

The first two authors would like to acknowledge the encouragement provided by their management for this book project. In particular, Prashant would like to thank Pei-Hsin Ho and Robert Damiano at the Advanced Technology Group in Synopsys, Inc., whereas Rupesh is grateful to Prashant Sawkar at the Enterprise Platform Group in Intel Corporation.

Prashant would like to thank his co-authors for successful collaborations not only on this book but also on several research projects. He is also obliged to Timothy Kam for inviting him to work with Rupesh on congestion, and to Xinning Wang for many valuable discussions. Much of his effort on this book was carried out during a five week "vacation" in India; he is grateful to his parents and in-laws for uncomplainingly sacrificing the time that they had expected to spend with him, when they instead found him focusing on this book. Their unflagging encouragement, along with the quest for learning instilled in Prashant by his parents and his doctoral advisor C. L. Liu, were instrumental in giving him the confidence to take up this book project. He could not have completed this project without the continual love and support of his wife Priti, who kept him well motivated (and well fed) in spite of the many lonely hours that he spent on the manuscript rather than with her.

Rupesh is grateful to the Strategic CAD Labs at Intel for offering him the opportunity to explore the area of congestion-aware logic synthesis and thus get acquainted with the routing congestion problem during summer internships in 2002 and 2003. He would particularly like to thank Xinning Wang, Timothy Kam, Steve Burns, Priyadarsan Patra, Michael Kishinevsky, and Brian Moore whose questions during these internships motivated the work leading to publications on technology mapping targeting routing congestion. He is also grateful to his co-authors: to Prashant, as this book is the result of a collaboration with him since 2002, and to Sachin, his doctoral advisor, for allowing him to continue to work on congestion-aware technology mapping at the University of Minnesota, Minneapolis. He would like to thank his family members in India, namely, his mother Keshar and siblings Sandhya and Tushar, who were quite supportive and encouraging during the course of the work on this book, which went on during evenings and weekends for more than a year.

Sachin would like to express thanks to his co-authors to appreciate the pleasure of working with them on this book.

Finally, the authors are grateful for the fantastic editorial support (and timely gentle prodding) provided by the staff at Springer Verlag. In particular, this book could not have been completed without the active involvement of Carl Harris and Katelyn Stanne.

Portland, OR
Hillsboro, OR
Minneapolis, MN

Prashant Saxena
Rupesh S. Shelar
Sachin S. Sapatnekar

November 2006

Contents

Part I

THE ORIGINS OF CONGESTION

1

AN INTRODUCTION TO ROUTING CONGESTION

A traditional standard cell design contains wires that implement the power supply network, clocks, and signal nets. All these wires share the same set of routing resources. With the number of cells in a typical design growing exponentially and the electrical properties of metal wires scaling poorly, the competition for preferred routing resources between the various interconnects that must be routed is becoming more severe. As a consequence, not only is routing congestion increasing, but it is also becoming more damaging to the quality of the designs.

Most conventional design flows synthesize the power supply and clock networks prior to the signal routing stage. The power supply and clock nets do not perform any logical operation, but provide crucial logistical support to the circuits that actually implement the desired logical functionality. The power supply network is designed accounting for several factors such as the current requirements of the design, acceptable bounds on the noise in the supply voltage, and electromigration constraints. This network is designed in the form of a grid which may or may not be regular. Typically, the power supply network is created first and has all the routing resources to choose from. The clock nets are routed next and still have relative freedom, since only the power supply grid has used up some of the routing resources when the clocks are being routed. The clocks, which synchronize the sequential elements in the design, have strict signal integrity and skew requirements. Although they are usually designed as trees in mainstream designs, high-end designs often use more sophisticated clocking schemes such as grids in order to meet their stricter delay and skew requirements (even though such schemes can consume significantly more routing resources). Furthermore, the clock wires are typically shielded or spaced so that the signals on the neighboring wires do not distort the clock waveform; the shielding and spacing also consume some routing resources. The signal nets are routed last and can only use the routing resources that have not been occupied by the power supply and clock wires. Therefore, these are the nets that face the problem of routing congestion most acutely.

In this chapter, we will first introduce the terminology used in the context of routing congestion in Section 1.1, reviewing the basic routing model along the way. Then, we will motivate the need for congestion awareness through a discussion of the harmful effects of congestion in Section 1.2. Next, in Section 1.3, we will try to understand why the problem of routing congestion is getting worse with time. Finally, we will lay out a roadmap for the rest of this book in Sections 1.4 and 1.5 by overviewing the metrics and the optimization schemes, respectively, that are used for congestion.

1.1 The Nature of Congestion

A design is said to exhibit routing congestion when the demand for the routing resources in some region within the design exceeds their supply. However, although this simple intuitive definition suffices to determine whether some design is congested or not, one has to rely on the underlying routing model in order to quantify the congestion and compare its severity in two different implementations of a design.

1.1.1 Basic Routing Model

The routing of standard cell designs follows the placement stage, which fixes the locations of all the cells in rows of uniform height(s) as shown in Fig. 1.1(a). In today's standard cell designs, there is usually no explicit routing space between adjacent rows, since the wires can be routed over the cells because of the availability of multiple metal layers. The entire routing space is tessellated into a grid array as shown in Fig. 1.1(b). The small subregions created by the tessellation of the routing region have variously been referred to as *grid cells*, *global routing cells*, *global routing tiles*, or *bins* in the literature. The dual graph of the tessellation is the *routing graph* $G(V, E)$, an example of which is shown in Fig. 1.2(a). In this graph, each vertex $v \in V$ represents a bin, and the edge $e(u, v) \in E$ represents the boundary between the bins u and v (for $u, v \in V$).

In the routing graph shown in Fig. 1.2(a), the vias and layers are not modeled explicitly. On the other hand, the graph in Fig. 1.2(b) explicitly models bins on two horizontal and two vertical layers, as well as the vias between the different routing layers. The horizontal line segments in this figure represent the boundaries of bins on the same horizontal routing layer, the vertical line segments correspond to vias between adjacent horizontal and vertical routing layers, and the remaining line segments denote the boundaries between the bins on the same vertical routing layer. The process of routing a net on such a graph, therefore, implicitly determines its layer assignment as well.

A routing graph that models each layer explicitly consumes considerably more memory than one that bundles all the layers together. It is possible to

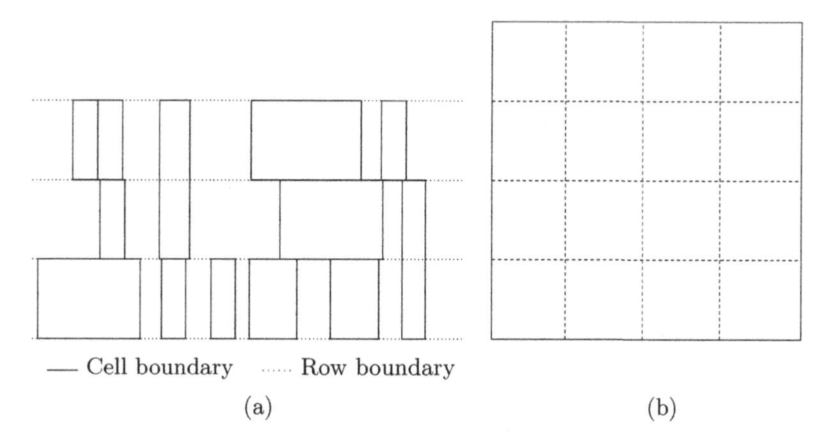

Fig. 1.1. (a) The placement of standard cells in rows of uniform height. (b) Tessellation of the routing area into bins.

retain most of the benefits of the layer-specific routing graph and yet reduce its memory footprint by grouping together layers that have similar electrical properties and wire widths. Today's process technologies offer up to nine metal layers [TSM04]. The lowermost one or two layers typically have the smallest wire widths and heights and are therefore the most resistive (although they can accommodate a larger number of tracks in each bin); these layers are appropriate for short, local wires. Minimum width wires on the middle routing layers are somewhat wider (and therefore, somewhat less resistive); these layers are used for the bulk of the global routing. The uppermost one or two layers are often reserved for very wide and tall wires that can provide low resistance paths for extremely critical signal nets and the global clock and power supply distributions.

Typically, most of the wires in a given layer are routed in the same direction (namely, either horizontal or vertical)[1], since this orthogonality of the layers simplifies the routing problem and allows for the use of the routing resources in a more effective manner. However, the lowermost layer is often allowed to be non-directional, since the flexibility to use both horizontal and vertical wires without needing a via between them facilitates pin hookups. Adjacent layers are usually oriented orthogonally to each other.

The bins are usually gridded using horizontal and vertical gridlines, referred to as *routing tracks*, along which wires can be created. The routers that use such grids are called *gridded routers*, whereas those that do not are said to be *gridless*. Although the use of the grid may appear to reduce the design freedom during routing, it allows for a simpler representation of the rout-

[1] Some process technologies do support diagonal wires in addition to the horizontal and vertical ones [XT03], but such routing architectures are not yet common in mainstream designs.

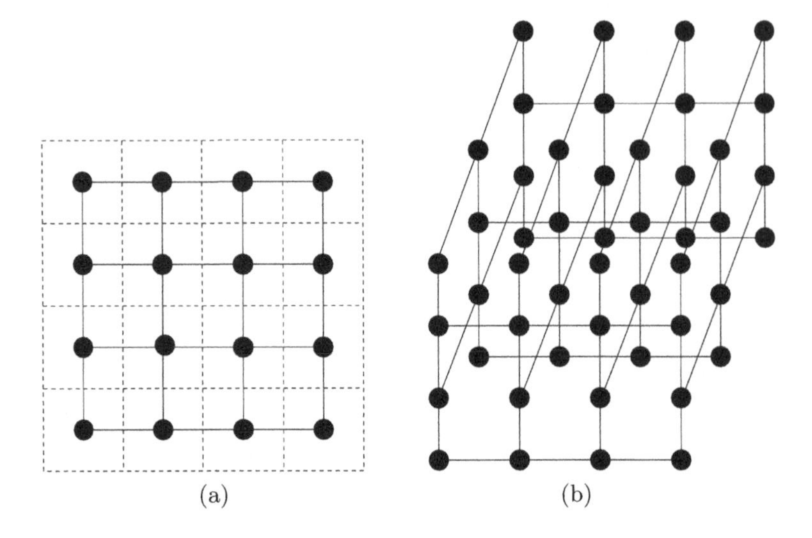

Fig. 1.2. (a) A routing graph that does not model vias and layers explicitly. (b) A routing graph for a four-layer routing architecture with explicit via and layer modeling.

ing configuration that can permit a more extensive exploration of the search space than would otherwise be possible. In particular, no special handling is required to handle via stacks across different layers, since the tracks in the grid automatically line up across the layers.

A bin can accommodate only a finite number of *routing tracks*, which may be contributed by several different layers if all the layers and vias have not been modeled explicitly in the routing graph. The number of tracks available in a bin denotes the *supply* of routing resources for that bin; this number is also known as the *capacity* of the bin. Similarly, the number of tracks crossing a bin boundary is referred to as the supply or the capacity of the routing graph edge corresponding to that boundary. A route passing through a bin or crossing a bin boundary requires a track in either the horizontal or the vertical direction. Thus, each such route contributes to the routing *demand* for that bin and edge.

Because of the complexity and the level of details involved in the routing, it is usually divided into two major stages, namely, *global routing* and *detailed routing*. The responsibility of the global routing stage is to generate routing topologies and embeddings for all the nets at the granularity of the bins. The layers in which a net is routed is determined by the layer assignment step. In today's routers, this step is usually performed simultaneously with the global routing. In other words, the regions through which a net is routed and the layers that it uses are determined concurrently. The subsequent detailed routing stage refines the global routing by assigning specific locations to all the wires within the bins, and legalizes the routing solution by eliminating routing design rule violations. Some routers use an explicit *track assignment* stage between the global and detailed routing stages [BSN+02], even though this step has traditionally been merged with detailed routing. Global and detailed routing algorithms are discussed in more depth in Sections 4.1 and 4.2, respectively, in Chapter 4.

A net N_i denotes the logical connectivity between its *pins* (also referred to as its *terminals*) that are located in some of the bins, and may be represented as $\{v_{i,1}, v_{i,2}, \cdots, v_{i,k_i}\} \subseteq V$, where k_i is the number of terminals of N_i. Usually, one[2] of the pins of the net represents its *driver* or *source*, and the remaining pins represent its *receivers* or *sinks*. For every net N_i, the objective of the global routing stage is to find an additional subset of vertices $V_{S_i} \subset V$ (referred to as its *Steiner* nodes) and a set of edges $E_i = \{e_{i,1}, e_{i,2}, \cdots\} \subset E$ in the routing graph so as to form a rectilinear spanning tree $T_i = (V_i, E_i)$, where $V_i = N_i \cup V_{S_i}$, that minimizes some cost metric (such as the total wirelength of the net, the delay to its most critical sink, or the maximum congestion along the route of the net). When the global routing of all the nets has been completed, the demand for routing tracks is known for each bin as well as for each bin boundary. One of the objectives of the global routing stage is to route all the nets such that the demand for tracks in any bin does not exceed the supply of the tracks in that bin.

The global routing stage is followed by the detailed routing. The detailed router typically handles small regions consisting of a few bins at a time, and focuses on generating a clean routing that does not violate any design rules. This is in contrast to global routing that operates on the entire routing area, and can abstract away many of the detailed design rules. The success of detailed routing depends heavily on the quality of the results obtained during the preceding global routing. For instance, if the global routing has assigned more nets to a bin than the number of available tracks, then the successful detailed routing of all the nets in that bin may not be possible. When more wires than can be accommodated on the tracks in a bin compete to pass through that bin (even after attempting to find alternative global routes for the nets routed through that bin through uncongested bins), the routing may remain incomplete with either opens or shorts on the wires, or some of these wires may be detoured. The occurrence of such a scenario is referred to as *routing congestion*; it hints at the unavailability of sufficient routing resources in particular regions to successfully route the wires assigned to those regions.

Figure 1.3 depicts some of the steps involved in the routing of a net connecting terminals pin_1 and pin_2. The selection of the bins for the global routing of the net is shown in Fig. 1.3(b). Most detailed routers perform track

---

[2] A few nets can have multiple parallel drivers if the total capacitive load of the net and its sinks is too large for a single driver. Moreover, some nets (such as bidirectional buses) can have multiple drivers, at most one of which may be active at any time, with the remaining drivers being cut off using the high impedance state in tristate logic.

assignment either independently or along with the creation of the wire segments and vias in the final layout of the net. As shown in Fig. 1.3(c), the track assignment step chooses the tracks for a net, typically optimizing cost functions such as delay, number of vias, crosstalk noise, etc.. Finally, as shown in Fig. 1.3(d), the detailed router completes the layout by creating wires of appropriate length and by generating vias where required, while taking into account all the design rules.

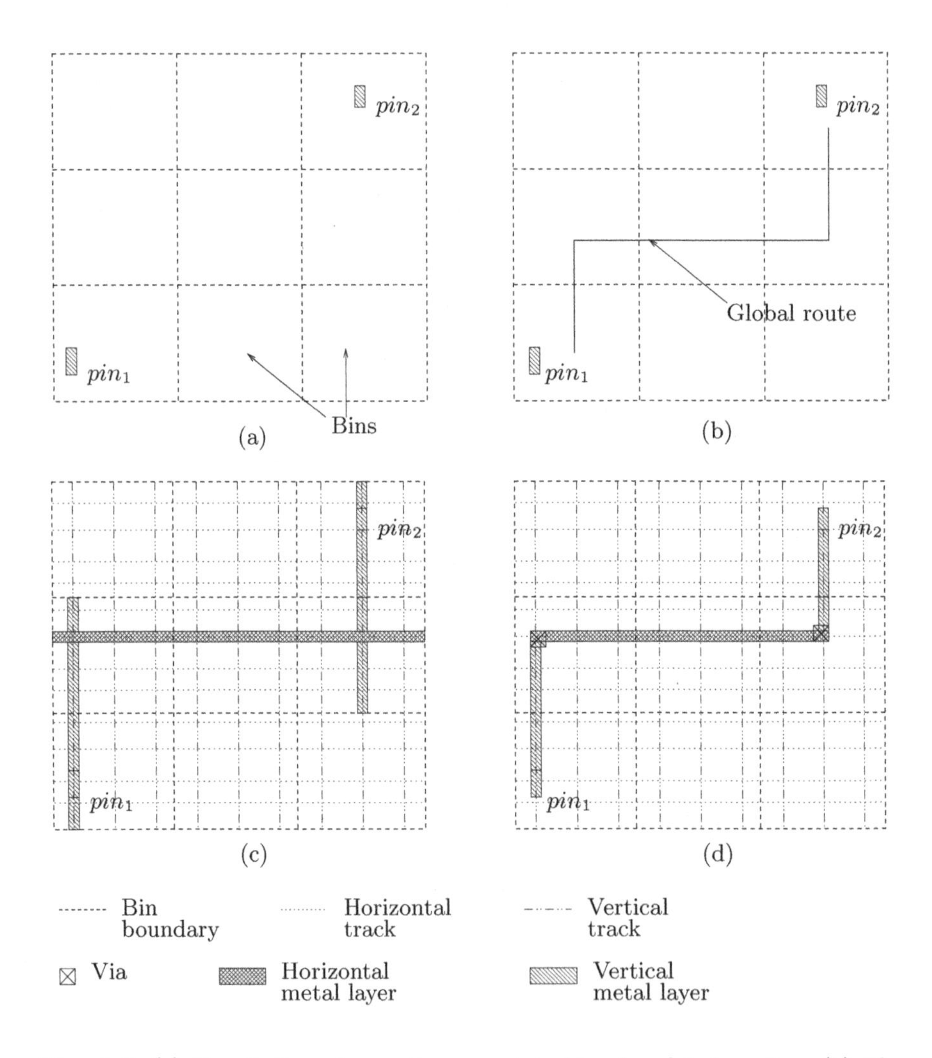

Fig. 1.3. (a) Tessellation of a routing area into bins for global routing. (b) The global routing of a net connecting pin_1 and pin_2. (c) Selection of horizontal and vertical tracks during track assignment. (d) Creation of the final routing that obeys all design rules.

1.1.2 Routing Congestion Terminology

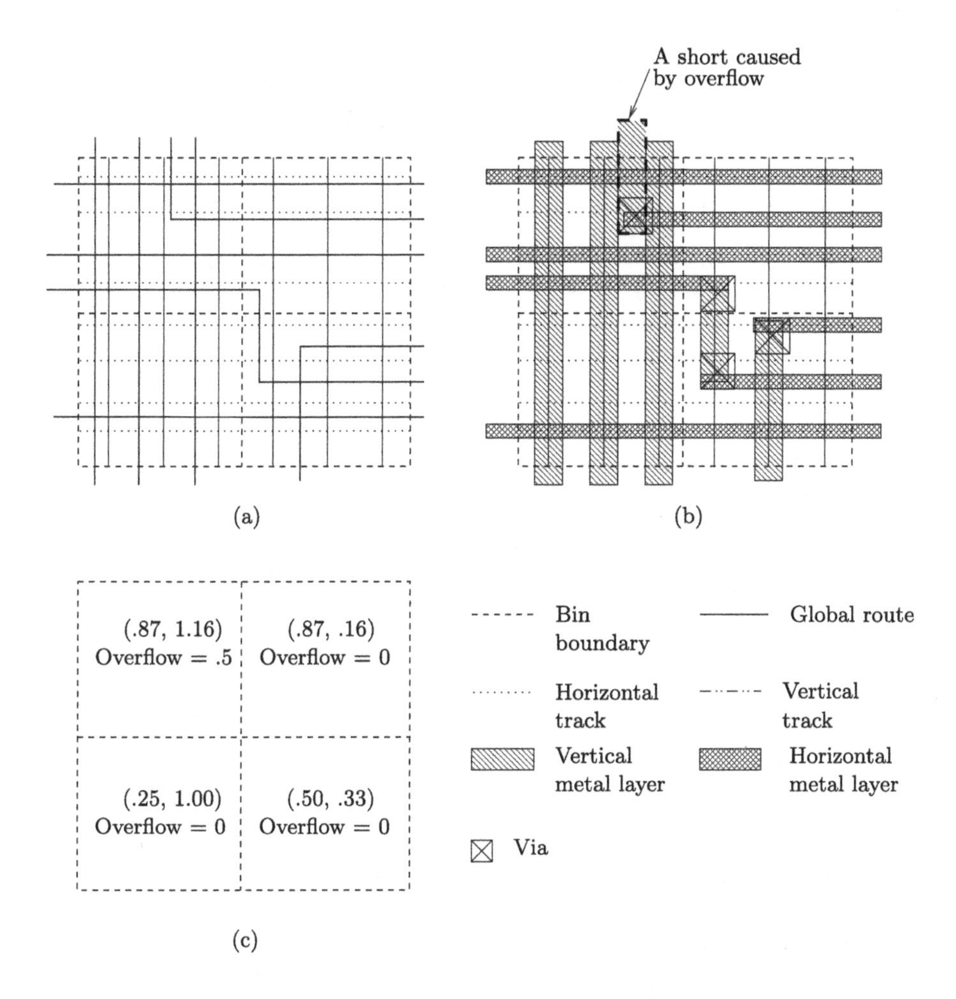

Fig. 1.4. (a) An example of a global routing. (b) A corresponding detailed routing showing a short in an overflowing bin. (c) The congestion and track overflow in each bin.

The existence of routing congestion is often manifested as detoured wires, poor layer assignment, or incomplete routes containing opens and shorts. As an example, consider Fig. 1.4, which depicts nine nets routed through four adjacent bins. Let us assume that each bin in the figure accommodates three vertical tracks and four horizontal tracks. The global routing of the nets is shown in Fig. 1.4(a), where one can observe that four nets are assigned to vertical tracks in the top left bin. Since this bin can accommodate only three vertical tracks, one of the nets there cannot be detailed routed successfully unless it is rerouted through some other bin. If no rerouting can be found, it may create a short, as shown in Fig. 1.4(b). The remaining bins do not show any routability problems, since the demand for horizontal or vertical routing tracks never exceeds their supply in any of those bins. In other words, all the bins in the figure except for the top left one are uncongested.

One of the metrics commonly used to gauge the severity of routing congestion is the *track overflow* that measures the number of excess tracks required to route the wires in a bin. It can be defined formally as follows:

Definition 1.1. *The horizontal (vertical) track overflow T_x^v (T_y^v) for a given bin v is defined as the difference between the number of horizontal (vertical) tracks required to route the nets through the bin and the available number of horizontal (vertical) tracks when this difference is positive, and zero otherwise.*

In other words,

$$T^v = \begin{cases} \text{demand}(v) - \text{supply}(v), & \text{demand}(v) > \text{supply}(v), \\ 0 & \text{otherwise.} \end{cases}$$

Throughout this book, whenever the routing direction is left unspecified in some equation or discussion, it is implied that the equation or discussion is equally applicable to both the horizontal and the vertical directions. Thus, for instance, usage of the notation T^v (for the track overflow) in a statement implies that the statement is equally applicable to both T_x^v and T_y^v. In the same vein, if the bin to which a congestion metric pertains is clear from the context, it may be dropped from the notation (as is the case with T_x and T_y in the following paragraph).

If we assume that a route segment that enters a bin and then terminates inside that bin consumes half a routing track within that bin, it is easy to verify that $T_y = 0.5$, $T_x = 0$ for the top left bin, and $T_y = T_x = 0$ for all other bins in Fig. 1.4.

The formal definition of the *congestion* metric is as follows:

Definition 1.2. *The horizontal (vertical) congestion C_x^v (C_y^v) for a given bin v is the ratio of the number of horizontal (vertical) tracks required to route the nets assigned to that bin to the number of horizontal (vertical) tracks available.*

Thus, the congestion in a given bin is simply the ratio of demand for the tracks to their supply in that bin, and can be written as:

$$C^v = \frac{\text{demand}(v)}{\text{supply}(v)}.$$

Figure 1.4(c) also shows the horizontal and vertical routing congestion in each bin. The first element in each congestion 2-tuple associated with a bin denotes the horizontal routing congestion C_x, whereas the second represents the vertical congestion C_y. For instance, the bottom left bin has a congestion of $(0.25, 1.0)$, since the horizontal demand and supply for that bin are one and four routing tracks, respectively, whereas the vertical demand and supply are both three tracks each. One can observe that the top left bin has a congestion of 1.16 in the vertical direction, indicating that the demand for vertical routing tracks in that bin exceeds their available supply.

The overflow and congestion metrics can be defined similarly for the bin boundaries (or equivalently, for the routing graph edges). These definitions can also be further extended to consider each routing layer individually.

The notion of a *congestion map* is often used to obtain the complete picture of routing congestion over the entire routing area. The congestion map is a three-dimensional array of congestion 2-tuples indexed by bin locations and can be visualized by plotting congestion on z-axis while denoting bins on x- and y-axes. Such a visualization helps designers easily identify densely congested areas (that correspond to peaks in the congestion map).

Some other commonly used metrics that capture overall routability of the design rely on scalar values (in contrast to three-dimensional congestion map vectors). These metrics include the total track overflow, maximum congestion, and the number of congested bins, and are defined as follows:

Definition 1.3. *The total track overflow (OF) is defined as the sum of the individual track overflows in all of the bins in the block.*

In other words,

$$OF = \sum_{\forall v \in V} T^v.$$

Definition 1.4. *The maximum congestion (MC) is defined as the maximum of the congestion values over all of the bins in the block.*

In other words,

$$MC = \max\{C^v : \forall v \in V\}.$$

Definition 1.5. *The number of congested bins (NC) is defined as the number of bins in the block whose congestion is greater than some specified threshold $C_{threshold}$.*

It can be written as:

$$NC = |\{v : v \in V \text{ and } C^v > C_{threshold}\}|$$

Note that the number of congested bins in a design is a function of the value of the selected threshold congestion. The designer may choose $C_{threshold}$ to be 1.0 to find the number of bins where the demand exceeds the supply, or may select some slightly smaller value (such as 0.9) to identify the bins where the nets are barely routable.

1.2 The Undesirability of Congestion

Designers attempt to minimize the routing congestion in a design because congestion can lead to several serious problems.

- It may worsen the performance of the design.
- It may add more uncertainty to the design closure process.
- It may result in degraded functional and parametric yield during the manufacturing of the integrated circuits for the design.

We discuss each of these issues in detail in the following subsections.

1.2.1 Impact on Circuit Performance

With wire delays no longer being insignificant in modern process technologies, an unexpected increase in the delay of a net that lies on a critical path can cause a design to miss its frequency target. The most common reason for such an unexpected increase in the delay of a net is routing congestion. Congestion can affect the delay of a net in several ways (that are listed next and then elaborated upon in the remainder of this section):

- The routing of the net may be forced to use the more resistive metal layers, resulting in an increase in the delay of the net.
- The routing of the net may involve a detour created to avoid passing through congested regions. This detour will increase the delay of the net as well as that of its driver.
- The routing may include a large number of vias generated when the router attempts to find a shortest path route through (or complete the detailed routing in) a congested region containing numerous obstructions corresponding to the nets routed earlier, leading to an increase in the delay of the net.
- Wires routed in a congested region may be more susceptible to interconnect crosstalk, leading to a greater variation in the delays of the nets.

A good timing-driven global router will attempt to route long or timing-critical nets on the less resistive upper layers, where the improved wire delays can amortize the via stack penalties involved in accessing those layers. However, if those preferred layers have already been occupied by other nets (that are presumably also critical), then the lower layers that are usually more resistive may also have to be used for some of the critical nets that are routed later. The resulting increase in the delays of the critical nets routed on the lower layers can cause timing violations on the paths passing through those nets.

If a net is detoured to avoid a congested region, the detour can increase the delay not only of the net but also of its driver. Even if we use a very simple (lumped parasitic) delay model, it is easy to show that the delay of an unbuffered net increases quadratically with its length. Under this delay model, the delay $D_w(l)$ of a wire of length l can be written as:

$$D_w(l) = (rl)(cl) = rcl^2, \tag{1.1}$$

where r and c are, respectively, the per unit length resistance and capacitance of the wire. In other words, an increase in the wirelength of a net caused due to a detour results in its delay growing quadratically with that increase. (A similar relationship can also be shown using more sophisticated delay models). Furthermore, the increased wirelength also raises the total capacitive load seen by the driver of the net, increasing its switching time. This additional capacitance also results in an increase in the dynamic power dissipation in the net. If the driver of the net needs to be sized up to drive the increased wire load, or if the detour is large enough to require the insertion of buffers, the leakage power may also increase.

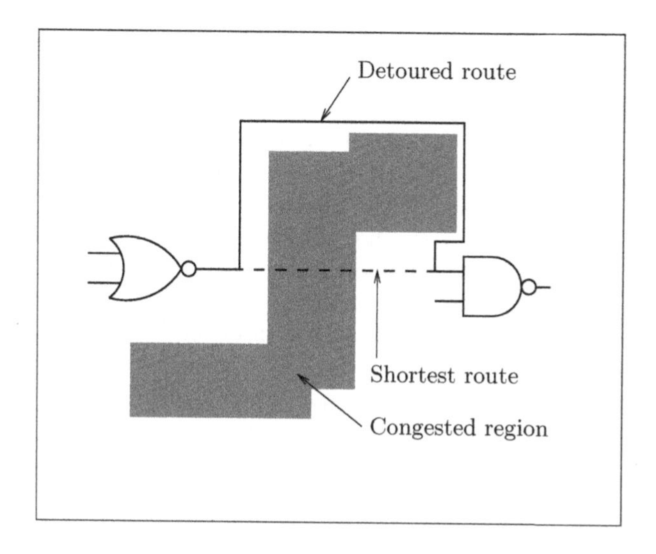

Fig. 1.5. A wire detoured because of congestion.

For example, Fig. 1.5 illustrates a scenario in which a congested region forces a net to detour significantly. Let us assume that the length of the shortest possible route for the net is 300μ, whereas that of the actual route is 700μ. Using representative values of $1.6\Omega/\mu$ for the resistance and $0.2fF/\mu$ for the capacitance of the wire, the delays of the wire based on the shortest possible and actual routes (as per Equation (1.1)) are $300 \times 1.6 \times 300 \times 0.2$ $ps = 28.8$ ps and $700 \times 1.6 \times 700 \times 0.2$ $ps = 156.8$ ps, respectively. Thus, in this case, even if we ignore the resistance of the larger number of vias that the detoured routing is likely to require, the delay of the net increases by a factor of more than five because of its detour.

The resistance of vias is scaling across process generations even more poorly than the resistance of wires. Vias in modern process generations can be significantly resistive, often being equivalent in resistance to as much as several tens of microns of minimum width wiring on one of the middle layers. Each via inserted into the routing of a net adds a significant resistance to that net, thus increasing its delay. Furthermore, most process technologies use *landed* vias for purposes of manufacturability; these vias are wider than the corresponding minimum width routing tracks. Therefore, these vias present additional blockages to the router, worsening the congestion even further.

In all the above cases, a secondary effect that can further aggravate the critical paths passing through nets that obtain a poor routing due to congestion is the worsened delays of the logic stages downstream from these nets. The increased resistance of these nets causes a degradation in the transition times for the rising and falling signal edges at their sinks. Consequently, the cells at the sinks of these nets also slow down[3].

Another potential problem that is aggravated in congested regions is that of interconnect crosstalk. A signal switching in a net driven by a strong driver can affect neighboring victim nets significantly. This interaction may result in a functional failure (if the coupled noise causes the logic value stored in a sequential element or at the output of some non-restoring logic element such as a domino gate to flip), or in a widening of the switching windows in the neighboring nets. The latter effect leads to an increased variation in the delays of the victim nets, because their effective capacitance varies depending on the switching state of their neighboring aggressor nets. Although gate sizing and buffering can ameliorate some of the noise problems, other instances of these problems are best fixed through the insertion of shields between the aggressor and victim nets, or by spacing the victim nets farther away from their aggressors. However, these techniques are difficult to apply in congested layouts because of a shortage of routing resources.

1.2.2 Impact on Design Convergence

Routing congestion adds unpredictability to the design cycle. This unpredictability of design convergence can manifest itself in two ways (that are discussed in the remainder of this section):

- Congestion-oblivious net delay estimates may mislead the design optimization trajectory by failing to correctly identify the truly critical paths.
- If a block cannot be successfully routed within its assigned area in a hierarchical design flow, the block designer may need to negotiate with the designers of neighboring blocks for more space, thus possibly necessitating a redesign of those blocks also.

[3] Note, however, that this dependence of the delay of a cell on its input slews is not captured by first-order delay models.

If the effects of congestion such as detours, unroutability and the selective delay degradation of some nets are not adequately modeled during logic synthesis or physical synthesis, then the optimizations applied at these stages can be easily misled by erroneous estimates for the design metrics. This may lead the optimization trajectory to poor configurations in the design space, recovering from which may require design iterations that are not only time-consuming but also may not guarantee convergence.

As an illustration, if the placement is oblivious to the routing congestion, the placer will not be able to position the cells so that the critical nets avoid congested regions. This also has impact on the timing, as can be seen from the example in Fig. 1.5. In this example, the placer may no longer try to position the two depicted cells any closer if the net delay of 28.8 *ps* (computed in Section 1.2.1) that it has estimated using the shortest route assumption is acceptable, even though the actual delay of 156.8 *ps* results in a timing violation.

As another example, consider the case of two nets N_1 and N_2 such that the former lies in a very densely congested region that causes its actual delay to be several times larger than its estimated delay, whereas the latter has full access to preferred routing resources. Furthermore, let us assume that N_2 is slightly more critical than N_1 on the basis of the estimated net delays. In this scenario, a circuit optimization engine that does not comprehend routing congestion will select a path through N_2 for optimization, even though the actual criticality (*i.e.*, the criticality based on achievable net delays rather than estimated ones) of some path through N_1 may be considerably higher.

While computing net parasitics and net delays, several of today's placement engines use a length-layer table that attempts to mimic how a designer or a good performance-driven router would ideally assign the layers to the nets based on their lengths. For instance, long and timing-critical nets would be assigned to upper layers, whereas short and non-critical nets would be allocated to the lower layers. However, the wire delays based on such a table quickly become invalid in congested designs, when the router is unable to route nets on the layers to which they have been assigned by the placer because of congestion. Again, this mismatch between the assumed and actual layers on which a net is routed can invalidate many of the optimizations applied to the net during physical synthesis.

Although the layer assignment and detour assumptions for a given net can be enforced using the rip-up and reroute of other nets in its vicinity or by changing the net ordering, this procedure may merely cause some other nets to become critical because of their poor routes. Indeed, if the unexpected detours are large, the nets may require buffering or significant driver upsizing. Buffer insertion can aggravate the congestion because of the extra via stacks required to access the buffers and the reduced flexibility in rerouting buffered nets. Furthermore, both newly inserted buffers and upsized drivers can create cell overlaps, whose resolution through placement legalization can cause more nets to be rerouted, often invalidating their assumed delays in the process.

If a design is unroutable or fails to converge timing because of unexpected net delays, the design flow may need to revisit the global placement stage in the hope of generating a more routable placement. If the placement stage is not able to resolve the routability issue, it may become necessary to remap or resynthesize either some or all of the logic. These additional design iterations are not only expensive from a runtime perspective, but, more significantly, may result in layouts whose congestion profile is no better than that obtained in the original iteration, unless the design flow is congestion-aware.

The unroutability of a block can prove particularly problematic in the hierarchical design methodologies that are used for large, complex designs. In these designs, different blocks are independently designed in parallel by different designers. This ability to design the blocks in parallel crucially depends on the designers obeying mutually negotiated physical, temporal, and logical interfaces for all the blocks. In high performance designs that use today's process technologies, blocks containing even a few tens of thousands of cells may require more than 30% white space in order to ensure routability. In such a scenario, the area required to route all the wires internal to a block is very difficult to predict accurately without actually implementing the block[4]. Therefore, late changes to the floorplan may be inevitable. If a block is found to be unroutable within the area allocated to it in the floorplan for the design, the need to increase its area or change its aspect ratio in order to accommodate the routing of its nets breaks the clean interface between the blocks, possibly requiring the redesign of its neighboring blocks that may already be in an advanced stage of implementation.

An example of this phenomenon is illustrated in Fig. 1.6, in which the area for block P is expanded to include the shaded area. This expansion occurs at the expense of the areas for its neighboring blocks Q and R, which now have to ensure that they remain routable in their newly reduced areas while still meeting all their design constraints.

When floorplan changes occur, all the affected blocks have to be resynthesized and laid out with new pin locations and modified block areas. These blocks are typically still required to meet the same delay constraints as they did earlier, although some power goals for individual blocks may need to be recomputed in the process of redistributing the overall power budget among the blocks. Converging the design using the new floorplan presents a variety of challenges for both types of blocks – the ones whose areas have grown, as well as the others whose areas have shrunk. For example, the block P in the floorplan in Fig. 1.6(b) may have longer wires, on the average, than those for the same block placed in the smaller area shown in the original floorplan in Fig. 1.6(a). The additional capacitance of the longer wires results in larger

[4] Overestimating the white space for a block is undesirable, because it leads to an unnecessary increase in die area, which in turn increases the manufacturing cost and reduces the yield for the integrated circuits implementing the design. This will be expanded upon in Section 1.2.3.

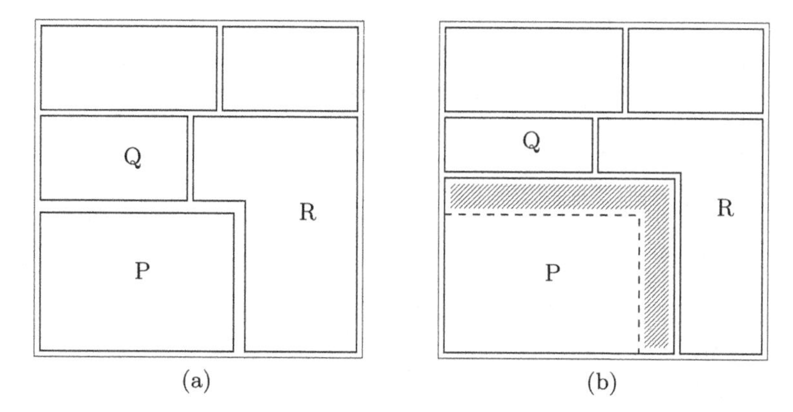

Fig. 1.6. (a) Original floorplan. (b) Altered floorplan to alleviate routing congestion in P by creating more space for the block P at the expense of the areas reserved for the blocks Q and R.

net delays, which in turn leads to increased power as drivers are upsized or extra buffers are added to the design. The blocks Q and R in the new floorplan face the challenge of placing and routing their logic in areas that are smaller than those allocated to them in the original floorplan. If either of the blocks Q or R cannot be converged in the new floorplan, more changes to the areas and shapes of neighboring blocks may be required, the success of which will not be known until those blocks have also been taken through entire synthesis-to-layout flow.

Thus, ensuring the routability of the entire design through floorplan changes involves a large cost, since it may involve many time-consuming iterations. While some of these iterations may be required anyway for delay and power budgeting across the blocks, routability adds one more factor that can necessitate additional iterations during the process of design convergence.

1.2.3 Impact on Yield

A densely congested design is likely to result in a lower manufacturing yield than a similar uncongested design. The yield of the integrated circuits implementing a design is affected by the congestion of the design in three ways:

- Congestion typically results in an increased number of vias in the routes, which can affect the yield.
- Congested layouts tend to have larger critical areas for the creation of shorts and opens due to random defects.
- Any increase in the area of a congested design in order to accommodate the routings of all its nets typically leads to some yield loss.

As mentioned in Section 1.2.1, routes in congested regions typically contain a larger number of vias than similar routes in uncongested regions. When a router tries to find a shortest path route for a net passing through a congested region, it may create numerous bends when it maneuvers around the obstructions created by other routings. Each of these bends results in additional vias. Nets whose routes have been detoured in order to avoid congested regions also tend to have more vias than nets that have more direct routes passing through uncongested regions. Furthermore, during the process of detailed routing, if a local region is heavily congested, the reduced flexibility of the router often results in the routes having to change layers frequently in the process of being maze routed to their pins (because direct routes with few layer changes may not be possible). This too leads to an increase in the number of vias required for the routing.

The existence of a large number of vias is problematic for two reasons. Firstly, since vias are often wider than the minimum widths of the routing tracks on the corresponding layers, a large number of vias may create routing blockages that may further aggravate the congestion (and consequently result in the generation of yet more vias). Secondly, having a large number of vias can lead to yield loss during manufacturing. Vias have traditionally been undesirable from the manufacturability point of view because of the mask alignment problem, which occurs because masks for two different metal layers are required to align perfectly in order to create the vias as desired. Although advances in manufacturing technologies have reduced the severity of this problem, vias are still harder to manufacture than wire segments. Furthermore, with the shrinking geometries of modern process technologies, vias are often a factor in parametric yield loss (*i.e.*, a reduction in the maximum frequency at which the integrated circuit can operate). Ensuring perfect electrical connectivity through a via requires the metal deposition to go all the way down to the lower metal layer, without the creation of any void. However, this is not very easy to enforce in practice. A void within a via can increase its resistance dramatically, causing the delay of the net containing the via to also grow appreciably. If this net lies on a critical path, it may lead to a parametric yield loss for the integrated circuits implementing the design.

Many industrial detailed routers try to minimize the potential parametric yield impact of vias by automatically inserting redundant vias in the routes wherever possible. (This also has the collateral benefit of reducing the effective resistance of the vias, leading to better net delays). However, this technique is not very applicable in the congested regions of the layout, where there may be little or no space available to insert additional vias.

The *critical area* of a layout is a metric that indicates the likelihood of a random defect particle of a given size to cause an open or short in the layout [Fer85, MD83]. Most of the wires in a congested layout are forced to be routed with no more than minimum spacing between them. This increases the critical area of the layout with respect to shorts, because a small deposition of extra metal between two neighboring wires can cause a short, leading to circuit failure and yield loss. In a sparsely congested layout, the wires can have larger spacings between them, reducing the possibility of such shorts. The critical area with respect to opens depends primarily on the total wirelength of the design (under the simplifying assumption that most of the signal nets are routed using minimum width wires, as is usually the case in practice). We have seen that routing congestion can result in an increase in wirelength because of detoured routes. Therefore, the probability of a random dust particle leading to an open on a wire also increases in such layouts.

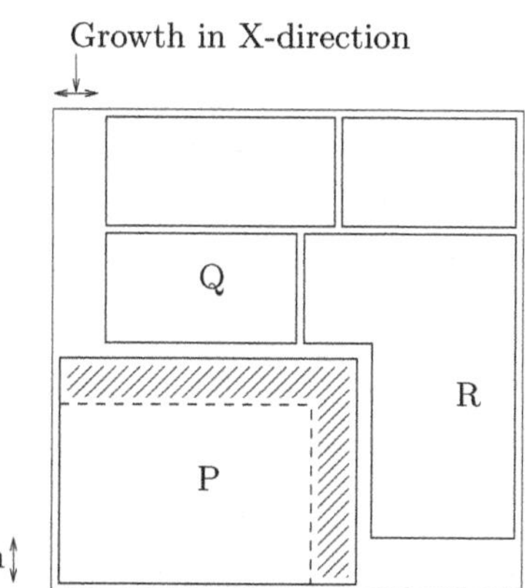

Fig. 1.7. Die size increase due to routing congestion in the block P.

Another way in which the yield can be affected due to congestion is through growth in the die size. As mentioned earlier, congestion can lead to a design being unroutable within its assigned area, if its nets cannot be routed completely even after the application of routing, placement, and synthesis optimizations targeted towards congestion minimization. In such a case, the die size may have to be increased to spread the cells out and make a larger number of routing tracks available, as depicted in the example in Fig. 1.7. In this figure, let us assume that the additional routing space required for the block P cannot be obtained at the cost of the block areas for Q and R. As a result, the die area for the chip is increased in both the horizontal and vertical directions to accommodate the growth of the block area for P. This increase in die area may, however, affect the yield adversely.

Many studies in the past have shown that the yield decreases with any increase in the die area, since the probability of random defects affecting the functionality of the circuit increases with the area. Several empirical models have been proposed to capture the relationship between the yield and die area, a typical example (based on a Poisson distribution for the occurrence of defects) [War74, Ber78] being:

$$Y = Y_0 e^{-AD},$$

where Y is the yield for chips whose die size is A, D is the defect density that depends on the clean room standards and the manufacturing process, and Y_0 is a constant. The above equation emphasizes the fact that any linear growth in area may affect the yield exponentially. Similar conclusions may also be drawn from most other yield models proposed in the literature. Thus, any increase in the die area due to routing congestion usually decreases the yield. A reduced yield translates into a rise in the per unit cost of the manufactured integrated circuits.

1.3 The Scaling of Congestion

We have seen that the existence of routing congestion in a design can lead to several serious problems. Unfortunately, an implication of today's design and process technology scaling trends is that the routing congestion problem will become even more acute in the coming years. In this section, we will build some intuition for this poor expected scaling of the routing congestion problem.

1.3.1 Effect of Design Complexity Scaling

The primary reason for the increase in congestion with successive process generations is design size scaling. With transistors getting smaller and cheaper with each successive process generation, it becomes feasible to pack more of them in a single design. Moore's Law, whose commonly accepted version states that the number of on-chip transistors is doubling every eighteen[5] months, is likely to continue to hold at least for the next decade (even if the rate of doubling slows down) [Moo03]. The semiconductor industry has managed to obey this law for the last four decades, enabling designers to integrate yet more functionality in their designs at each successive process node.

However, with an increase in the number of transistors and cells in a design comes a corresponding increase in the number of interconnections between them. Furthermore, the routing complexity of the designs also increases with an increase in the sizes of the designs. This can be illustrated by a simple thought experiment. Consider an optimized design in some process generation that is then shrunk to the next process generation without any change in logic or layout. This shrink involves the reduction of each of the geometric features of the original design by some scaling factor (that is typically $0.7\times$); these features include the sizes of all its gates as well as the widths, spacings, and lengths of all its wires. If one ignores the additional buffers that will be required to re-optimize the design at the new process node, one can argue that the

[5] Moore's original observation and prediction in 1965 was that the number of components in an electronic design was doubling every year [Moo65]. He later updated his predicted rate of doubling to once every two years in 1975 [Moo75].

routing complexity of the design has remained unchanged across the shrink; the insertion of additional buffers can only worsen the routing complexity. This is illustrated in Fig. 1.8, where the design shown in Fig. 1.8(a) is shrunk to the block placed in the lower left corner of the design depicted in Fig. 1.8(b).

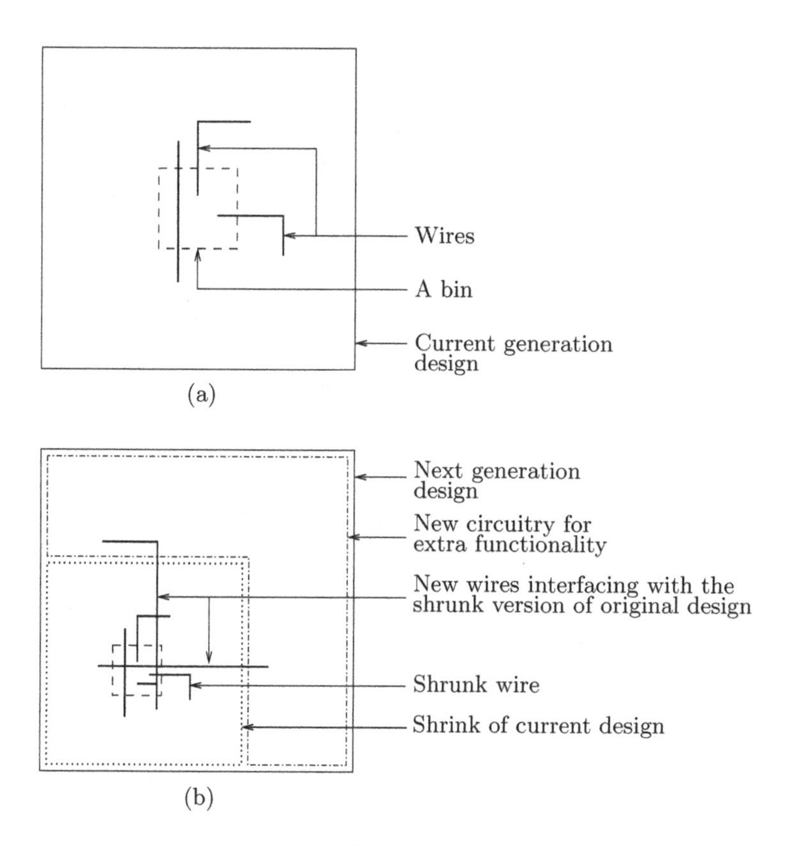

Fig. 1.8. (a) A bin with wires causing routing demand for a design in a given process technology. (b) The corresponding routing demand in a scaled version of the design that includes the original design as a shrunk block.

However, the new process node allows for the use of many more transistors, in accordance with Moore's Law. For the sake of simplicity, let us assume that the die size of the design and its cell density have remained constant[6] across the shrink. Therefore, the new design can accommodate twice as many transistors as in the old design; these additional transistors can be

[6] In practice, the die area usually grows slightly or remains unchanged across successive process generations, while the cell density decreases slightly in order to permit the successful routing of the wires in the shrunk design. As an example, the die sizes for recent Intel microprocessors have grown at the rate of 14% every two

used to integrate additional functionality, as shown in Fig. 1.8(b). However, the communication between this added functionality and the original block, which had earlier been implemented through off-chip interconnections (or not implemented at all), must now be through on-chip wires in the integrated design. These wires are in addition to the original on-chip routing of the shrunk block, and add to the routing complexity and congestion of that block. This increase in routing complexity is illustrated in Fig. 1.8 using the example of the wires passing through or connecting to a typical bin in the original design block.

This simple example illustrates the increase in routing complexity resulting from design size growth. One approach to handling this increased routing complexity could be through the introduction of new metal layers to accommodate the additional wires. However, each new metal layer involves significant additional mask generation costs. Furthermore, as will be discussed in Section 1.3.2, the introduction of additional metal layers is a strategy of diminishing returns in terms of easing routing congestion, in spite of the apparent extra routability afforded by these new layers. Consequently, the introduction of new metal layers has not been keeping pace with the rate at which the routing demands have been growing. Indeed, the number of routing layers has grown at the average rate of one new layer every three years over the last three decades, even though the number of transistors (and nets) in a design has been doubling every two years during this period.

Thus, worsening routing congestion is one of the costs that designers must pay in order to benefit from the increased integration made possible by process scaling. As design sizes increase, the routing congestion in those designs also becomes more severe. Since the introduction of additional routing layers does not help much in alleviating this congestion, designers tend to decrease the cell density and introduce more white space into their designs with each successive process generation in order to accommodate the increased routing demands.

1.3.2 Effect of Process Scaling

Although design size scaling is the prime reason behind the worsening of the routing congestion across successive process generations, the poor scaling of wires also plays a significant role in aggravating this problem. *Ideal* technology scaling [DGY+74] [Bak90] refers to the reduction of each dimension of the wires and the devices in a design by a constant shrink factor s (that has traditionally been $0.7\times$) while migrating a design from one process generation to the next. It is illustrated for wires in Fig. 1.9, where each dimension of the wire, *i.e.*, its width W, height H, distance D between the wires in the same layer, and interlayer dielectric thickness T shrinks by the constant scaling factor of s.

years [Bor00], whereas the number of transistors in the processors have doubled every two years during the same period.

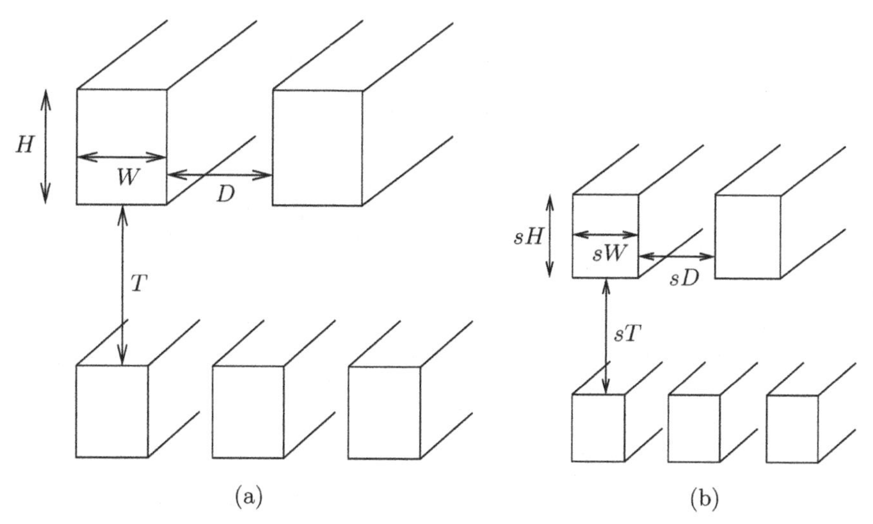

Fig. 1.9. (a) Interconnects in a given process technology. (b) Ideally scaled interconnects in the next process generation.

The resistance r, line-to-ground capacitance c, and the coupling capacitance c_c for a unit length of the wire in the current process technology generation are given by:

$$r = \frac{\rho}{W \times H},$$
$$c = \frac{\epsilon \times W}{T},$$
$$c_c = \frac{\epsilon \times H}{D},$$

where ρ is the resistivity of the metal (that was historically aluminum, but is usually copper in modern processes) and ϵ is the permittivity of the insulator, which is typically silicon dioxide. The corresponding quantities for the next process generation, where the dimensions of the wires are scaled as shown in the figure, are given by:

$$r^{next} = \frac{\rho}{sW \times sH},$$
$$c^{next} = \frac{\epsilon \times sW}{sT},$$
$$c_c^{next} = \frac{\epsilon \times sH}{sD}.$$

The above equations imply that the per unit length resistance of the wire doubles in each process generation, whereas the per unit length capacitances of the wire remain unchanged, as shown below:

$$
\begin{aligned}
r^{next} &= \frac{r}{s^2} = \frac{r}{0.7 \times 0.7} \approx 2r, \\
c^{next} &= c, \\
c_c^{next} &= c_c.
\end{aligned}
$$

As a result, assuming no activity on the neighboring wires, the delay $D_w^{next}(sl)$ for a wire of length sl obtained by shrinking a wire of length l (whose delay is denoted by $D_s(l)$) is given by the following equation:

$$
\begin{aligned}
D_w^{next}(sl) &= r^{next}(sl) \times (c^{next} + c_c^{next})(sl) \\
&= \frac{r}{s^2}(c + c_c)(sl)^2 \\
&= D_w(l).
\end{aligned}
$$

The above equation corresponds to the scaled delay of a local interconnect whose length shrinks by the usual shrink factor s. It indicates that in spite of the reduction in length, the delay of the wire does not decrease. This is in sharp contrast to the delays through the transistors, that typically speed up by a factor of s with every process generation.

The situation for global nets, whose length does not shrink with scaling (because the die size does not shrink), is even more dire. The delay of a global interconnect of length l is given by:

$$
\begin{aligned}
D_w^{next}(l) &= r^{next}l \times (c^{next} + c_c^{next})l \\
&= \frac{r}{s^2}(c + c_c)l^2 \\
&= \frac{D_w(l)}{s^2}.
\end{aligned}
$$

In other words, the delay of a global net doubles from one process generation to the next. Even with optimal buffering, it can be shown that the delay of these nets degrades by a factor of $\sqrt{s}$. Furthermore, the inter-buffer separation in an optimally buffered wire shrinks much faster than the geometric shrink rate (shrinking instead at the rate of $s\sqrt{s}$ [Bak90]), resulting in a rapid increase in the number of buffers inserted into the nets [SMC+04] (along with its ramifications on congestion).

Consequently, wire delays become increasingly dominant with every process generation. Furthermore, since these delays do not scale well as shown above, much of the expected benefit of obtaining faster circuits on scaled process nodes is lost. Therefore, process designers often use non-ideal scaling on the wires in order to make them less resistive and improve their delay. This is done by making them wider or taller than would be indicated by the ideal scaling recipe; this is referred to as the *reverse scaling* of wires [SK99]. However, this has other undesirable side effects:

- When the wires are made wide, the number of tracks available in a given area decreases in a proportionate manner.

- When the wires are made relatively tall, the coupling component of the capacitance increases, which in turn results in increased crosstalk noise on the interconnects, causing high uncertainty in timing and possible functional failures.

Furthermore, tall or wide wires are also power hungry because of their increased capacitance; not only do they result in increased switching power, but they also require larger drivers.

The widening of the wires directly affects the supply of routing resources, which in turn increases the routing congestion. In contrast, tall wires are susceptible to interconnect crosstalk. As discussed in Section 1.2.1, interconnect crosstalk not only widens the switching windows in the nets, but can also result in functional failures. Many of the techniques used to counter this problem, such as the insertion of shields or an increase in the spacing between the signal nets, also consume routing resources that may be in short supply in congested regions. On the other hand, noise optimization techniques such as buffer insertion create additional routing blockages because of via stacks. Furthermore, very tall wires with highly skewed aspect ratios are difficult to manufacture. Thus, making the wires either tall or wide in order to counter the poor scaling of the wires affects the supply of routing resources and worsens the routing congestion.

One could hope that the increase in routing congestion due to design size or process technology scaling may be countered by adding extra routing resources in the form of new metal layers. However, there are several reasons why the introduction of new routing layers is not a panacea for the routing congestion problem. The via stacks required to access the top few layers create significant blockages on each of the underlying layers. This can become an especially severe problem on the bottommost few layers, since the via stacks from all the layers lying above them create blockages on these layers. Furthermore, we have seen that the resistance of wires increases rapidly with each successive process technology generation, causing the delay of global wires to degrade severely even as the gates speed up with scaling. Although buffer insertion can help reduce the severity of this imbalance, these buffers, when inserted in nets routed on the upper layers, result in yet more via stacks and their consequent routing blockages.

Another consequence of the worsening resistance of the wires is that the metal usage by the power grid and the global clock distribution is growing rapidly in order to avoid excessive voltage droop and poor clock slews, delays and skews. Indeed, most of the tracks on the topmost one or two layers are often reserved largely for the global clock and power grid distributions along with a handful of the most critical global signal nets.

The rapid reduction in the feature sizes with each successive process technology generation makes it increasingly difficult to obtain high yields during integrated circuit manufacturing. As a result, there has been much work on developing the so-called *design for manufacturing* techniques in recent years.

However, even these techniques have their limitations. It is feared that they will no longer be sufficient to ensure adequate yields at the 32 nm technology node. Therefore, many semiconductor manufacturers have started developing *restrictive design rules* (RDRs) as a way of tackling this manufacturing challenge. The RDRs impose many restrictions on the layout configurations permissible for the devices and their local interconnections. These restrictions, in turn, may reduce the flexibility available to detailed routers, making it harder for them to achieve successful route completion in congested designs. Therefore, in the presence of the RDRs, it will become even more important to address any expected congestion problems up front before handing the design over to the layout tools.

Thus, as designs get larger and more complex and process technologies descend yet deeper into the nanometer realm, routing congestion will become even more severe a problem than it is today. Therefore, it will be important for design flows to be able to predict the existence of routing congestion in some region of the design as early as possible, and take meaningful optimization steps to alleviate it with minimal impact to the primary design metrics such as performance, power and area.

1.4 The Estimation of Congestion

Routing congestion can be measured accurately only after the routing has been completed. However, if the design exhibits congestion problems at that stage, mere rerouting of the nets may not be able to resolve these problems. This may necessitate a new design iteration with changes being made to the placement or the netlist. For those changes to be effective, the designer must be able to judge whether the modified design is likely to have an improved congestion profile after it has been fully routed. It is in order to make this judgment that several congestion estimation metrics and schemes applicable to different stages of the design flow have been developed over the years.

The congestion metrics, therefore, serve two purposes. They allow the designer to predict the final routability of a point in the design space at a given design stage without actually going through the entire downstream flow. Secondly, they can guide the optimization techniques at that stage to move the design point towards a more routing-friendly implementation. The expectations from the metrics for these two purposes are slightly different from each other. For the former goal, accuracy is paramount and long computation times may be tolerated. However, for the latter purpose, good fidelity may be sufficient, but the metric must be fast to compute, since it will be repeatedly used to choose between different implementation choices during the course of the design optimization.

Several metrics that serve these purposes, at different stages in the design flow, have evolved over the last few years. At the routing stage, a number of such metrics are defined on the congestion map. As we saw in Section 1.1, these metrics include the track overflow, the maximum congestion, and the number of congested bins. The goal of most of the congestion metrics developed for earlier stages in the design flow is to predict these routing-level congestion metrics as accurately as possible. All these metrics are the subject of Part II of this book.

At the placement stage, fast but relatively inaccurate congestion predictors such as the Rent's exponent, pin density, perimeter degree, and wirelength can guide the optimization in early iterations, whereas accurate and expensive techniques such as probabilistic congestion maps and fast global routers can be invoked once the placement has stabilized. These metrics and congestion estimation techniques are discussed in detail in Chapter 2. Some of these metrics and techniques have also been extended to be applicable at the preceding technology mapping stage, especially when that stage incorporates some placement information (as is the case with most modern physical synthesis flows). Other proxies for congestion that are targeted for use during technology mapping are independent of the placement and rely solely on the structural, graph theoretic properties of the netlist. Congestion metrics for the technology-independent logic synthesis stage rely almost exclusively on the structural properties of the netlist. Congestion estimation metrics applicable during technology mapping and logic synthesis are the subject of Chapter 3.

1.5 The Optimization of Congestion

The elimination of routing congestion in a typical design flow has traditionally been the responsibility of the routing stage. However, with the severity of the congestion problem increasing over the years, industrial tools have been forced to build congestion awareness in upstream design stages also. Modern congestion-aware physical synthesis flows usually model design routability at the placement stage, and use various heuristics to improve the estimated congestion profile of the design at that stage.

Indeed, routing congestion can be considered for optimization at various stages in the design flow, as each stage offers different flexibilities. For example, nets can be routed differently to avoid congested regions during the routing stage. While performing placement, cells can be placed so that the corresponding design has fewer and less severe congestion hot spots. The technology-independent logic synthesis and technology mapping stages determine the structural properties of the underlying network and the individual nets in the design, which are the sources of the routing demand. In general, as the level of the abstraction of the design increases, so does the design freedom. At any design stage, the designer has access not only to the flexibility at the current stage but also to those at subsequent stages. Unfortunately, the accuracy of the congestion metrics decreases with the increasing level of design abstraction. This affects the overall effectiveness of this approach of fixing potential congestion problems as early as possible, since this approach depends not only on the design flexibilities but also on the accuracy and fidelity of the congestion metrics. Part III of this book is devoted to the optimization techniques available at various stages in the design flow and their effectiveness in improving the routability.

In the past, routing techniques such as rip-up and reroute or route spreading using congestion-based cost functions often sufficed for route completion. These are the topic of Chapter 4, which is dedicated to routing techniques aimed at relieving congestion. This chapter also discusses recent developments in the congestion-aware optimization of critical nets, as well as the interaction between signal routing and the power grid. Recent years have seen a significant emphasis on congestion alleviation during the placement stage, since relying solely on routing techniques for this purpose has often proven time-consuming and unpredictable for many modern designs. Chapter 5 describes the most important of these placement techniques in detail. Recent physical synthesis offerings from commercial vendors permit limited logic transformations during the placement optimizations. These capabilities point partly to the limitations of the placement-only techniques while optimizing the layout of a design and partly to the effectiveness of the logic transformations when they are guided by accurate placement information. Congestion-aware technology mapping is one such logic transformation that has seen much research during the last few years. This research has resulted in a few promising techniques to alleviate routing congestion. Technology-independent logic synthesis targeting routability or wirelength has also been pursued, typically by employing graph theoretic metrics. Although the effectiveness of such a transformation is limited by the difficulty of predicting downstream congestion accurately at this stage, this continues to feed an active area of research. Chapter 6 covers the current state-of-the-art in congestion-aware technology mapping and logic synthesis optimizations. Finally, Chapter 7 briefly describes the impact of behavioral and architectural choices on the final congestion in a design.

1.6 Final Remarks

Although routing congestion manifests itself only at the very end of the typical synthesis-to-layout flow, it can lead to unacceptable design quality and lack of design closure. The surest way to avoid such unpleasant last minute surprises is to improve the predictability of the design flow. With placement already having been integrated with circuit optimization in modern physical synthesis flows, one of the biggest obstacles to improving this predictability is the behavior of the router on congested designs, that can lead to unexpectedly large wire delays for some of the nets. Therefore, it is certainly desirable to build congestion awareness into the optimization of a design.

In this chapter, we defined several metrics to capture various aspects of routing congestion. We then looked at the impact of the routing congestion on the performance, convergence, manufacturability, and yield of modern designs.

We observed that the routing congestion in future circuits is expected to be even more severe because of growing design complexity and continuing technology scaling. Finally, we motivated the need for congestion metrics at different stages in the design flow, along with optimization techniques that can utilize these metrics to help mitigate the congestion.

References

[Bak90] Bakoglu, H. B., *Circuits, Interconnections, and Packaging for VLSI*, New York, NY: Addison-Wesley, 1990.

[BSN+02] Batterywala, S., Shenoy, N., Nicholls, W., and Zhou, H., Track assignment: a desirable intermediate step between global routing and detailed routing, *Proceedings of the International Conference on Computer-Aided Design*, pp. 59–66, 2002.

[Ber78] Bernard, J., The IC yield problem: A tentative analysis for MOS/SOS circuits, *IEEE Transactions on Electron Devices* 25(8), pp. 939–944, Aug. 1978.

[Bor00] Borkar, S., Obeying Moore's law beyond 0.18 micron, *Proceedings of the International Conference on ASIC/SOC*, pp. 26–31, 2000.

[DGY+74] Dennard, R. H., Gaensslen, F. H., Yu, H.-N., Rideout, V. L., Bassous, E., and LeBlanc, A. R., Design of ion-implanted MOSFETs with very small physical dimensions, *IEEE Journal of Solid-State Circuits* 9(5), pp. 256–268, Oct. 1974.

[Fer85] Ferris-Prabhu, A. V., Modeling of Critical Area in Yield Forecasts, *IEEE Journal of Solid-State Circuits* SC-20(4), pp. 878–880, Aug. 1985.

[ITR05] International Technology Roadmap for Semiconductors, 2005 ed., available at `http://www.itrs.net/Links/2005ITRS/Interconnect2005.pdf`.

[MD83] Maly, W., and Deszczka, J., Yield Estimation Model for VLSI Artwork Evaluation, *Electronics Letters* 19(6), pp. 226–227, March 1983.

[Moo65] Moore, G. E., Cramming more components onto integrated circuits, *Electronics Magazine*, pp. 114–117, April 1965.

[Moo75] Moore, G. E., Progress in digital integrated electronics, *Proceedings of the International Electron Devices Meeting*, vol. 21, pp. 11–13, 1975.

[Moo03] Moore, G. E., No exponential is forever: but "forever" can be delayed!, *Proceedings of the International Solid-State Circuits Conference*, vol. 1, pp. 20–23, 2003.

[SMC+04] Saxena, P., Menezes, N., Cocchini, P., and Kirkpatrick, D. A., Repeater scaling and its impact on CAD, *IEEE Transactions on Computer-Aided Design of Integrated Circuits and Systems* 23(4), pp. 451–463, April 2004.

[SK99] Sylvester, D., and Keutzer, K., Getting to the bottom of deep submicron II: A global wiring paradigm, *Proceedings of the International Symposium on Physical Design*, pp. 193–200, 1999.

[TSM04] TSMC unveils Nexsys^{SM} 90-nanometer process technology, 2004, available at `http://www.tsmc.com/english/b_technology/b01_platform/b010101_90nm.htm`.

[War74] Warner, R. M., Applying a composite model to the IC yield problem, *IEEE Journal of Solid-State Circuits* 9(3), pp. 86–95, June 1974.

[XT03] X Initiative, First 90 nm functional silicon using X architecture, 2003, available at `http://www.xinitiative.org/img/Toshiba_100803_Yoshimori.pdf`.

Part II

THE ESTIMATION OF CONGESTION

2

PLACEMENT-LEVEL METRICS FOR ROUTING CONGESTION

The minimization of routing congestion has traditionally been carried out during the routing stage. However, given the severity of the congestion problem in many modern designs, it is often too late to resolve all congestion issues during routing if the previous stages of the design flow have been oblivious to congestion. Most industrial congestion-aware physical synthesis flows rely heavily on improving the routability of a design during the placement stage itself. Indeed, as will be discussed in Chapter 5, there have been numerous theoretical and practical advances in recent years towards building congestion awareness into most of the standard placement paradigms. The placement stage is particularly appropriate for congestion mitigation because it provides significantly more flexibility than the routing stage. At the same time, congestion gains realized during placement are unlikely to be adversely affected by any subsequent design optimization steps (in contrast to those obtained during, say, technology mapping, that may be frittered away if the subsequent placement is not congestion-aware), because placement is followed immediately by the routing stage.

However, for a placement algorithm to be congestion-aware, it must first be able to evaluate whether any given placement configuration is likely to be congested after routing, as well as discriminate between any two placement configurations based on their expected congestion. Although running a router can certainly provide these capabilities, it is not practical to route the entire design every time its congestion must be evaluated (which may happen repeatedly within the iterations of the placement engine). Therefore, in order to develop congestion-aware placement, one must also develop methods that can be used to predict the expected post-routing congestion in a design without having to incur the significant runtime penalty involved in routing. Many such metrics and techniques, representing different tradeoffs between accuracy and efficiency, have been developed over the years for application during placement. Indeed, some of these techniques can even be extended for use during congestion-aware technology mapping, as will be discussed in Chapter 3.

Different congestion metrics provide different tradeoffs between the computational overhead required for their estimation and the accuracy that they can provide. They range from "quick and dirty" proxies for congestion, such as the total wirelength of the design, to expensive but accurate congestion prediction techniques such as the use of a fast global router. The remainder of this chapter discusses the pros and cons of the various methods that have been proposed for congestion estimation at the placement stage. We begin with a discussion of some of the simpler metrics in Section 2.1. This is followed in Section 2.2 by a discussion of probabilistic methods for congestion estimation; these techniques provide a good tradeoff between runtime and accuracy and have proven valuable in several industrial tools. Finally, we explore the use of fast global routers for congestion estimation in Section 2.3. Although fast global routers are computationally the most expensive among all the techniques that we discuss in this chapter, the significant correlation between the predicted and actual congestion maps that is achievable with their use is making them a popular choice in several modern physical synthesis flows.

Although the congestion of a placed layout can be approximated by a scalar such as the total wirelength of a design, it is more informative to compute the congestion individually in every bin that will subsequently be used for global routing. This generates a two-dimensional congestion map, as was mentioned in Chapter 1. The congestion within any given bin arises from three[1] kinds of nets, namely, (i) *intra-bin* nets, (ii) *inter-bin* nets with at least one pin within that bin, and (iii) *flyover* nets which are inter-bin nets that are routed through the bin but have no pin within the bin. For instance, net n_1 in the example in Fig. 2.1 is an intra-bin net, whereas the remaining nets are inter-bin nets. Furthermore, net n_3 is a flyover net from the perspective of the central bin in the layout, as are n_2 and n_4 with respect to the rightmost bin in the middle row and the bottom bin in the middle column of the layout, respectively. Different congestion estimation techniques handle these three classes of nets in different ways. The prerouting estimation of the congestion caused due to the inter-bin nets, especially that caused by flyover nets, is considerably more difficult than the estimation of the congestion caused due to the intra-bin nets, since global nets may have several choices for their routings. Therefore, their contribution to the routing demand in a given bin may not be clear *a priori*.

2.1 Fast Metrics For Routing Congestion

In this section, we will discuss several fast but relatively inaccurate (or, "quick and dirty") metrics that have been used during global placement. These met-

[1] Unfortunately, there is no standardized taxonomy of nets across the literature. Some works, such as [HM02], refer to the nets with one or more pins in a given bin as the nets *local* to that bin (independent of their physical span or wirelength) and the ones with no pins in the bin as the *global* nets with respect to that bin.

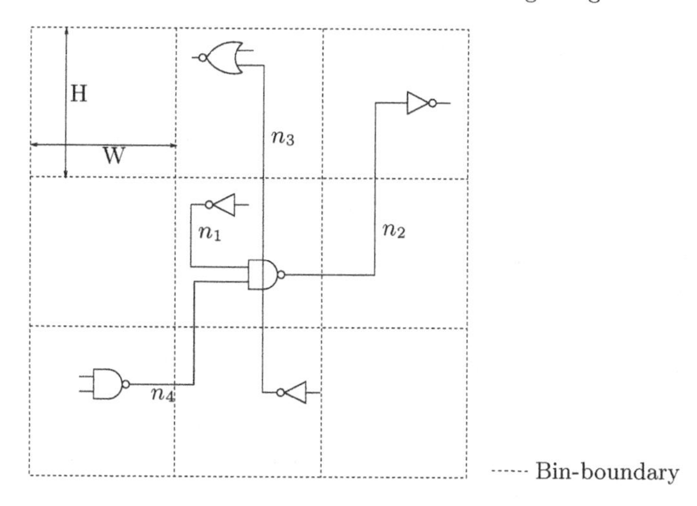

Fig. 2.1. Intra-bin and inter-bin nets.

rics range from scalars such as the total predicted wirelength of the design, to spatially distributed metrics that enable a quick estimate of the congestion expected in each bin, such as the *pin density* and the *perimeter degree.* We will also look at the application of certain structural properties of the circuit graph, captured by a relationship known as Rent's rule, to speed up the computation of the metrics discussed in this section even further.

These metrics are best used by fast congestion analyzers embedded within optimizers during the early stages of global placement. During these applications, their fidelity to the actual congestion can help choose between alternative optimization moves based on their expected congestion impact, without creating a significant computation overhead. Since these metrics do not capture the entire congestion in any given region, they are not very effective at determining whether a given placement is congestion-free, in order to decide whether the design flow may proceed to the routing stage. More accurate (but also more computationally expensive) metrics such as probabilistic or global routing-based congestion maps, which are described in later sections, are more appropriate to guide that decision process.

2.1.1 Total Wirelength

Traditionally, placers have targeted the minimization of cost functions involving wirelength in the belief that the optimization of the wirelength also leads to a reduction in congestion. Indeed, after performing the routing, the total wirelength (TWL) of a design can be written as the weighted sum of its congestion in each bin, summed over all the bins in its layout as follows:

$$\mathrm{TWL} = \sum_{n \in \mathcal{N}} l_n = \sum_{B} d(B) = \sum_{B} C^B s(B).$$

In the above equations, l_n denotes the length of the net[2] n, $d(B)$, $s(B)$ and C^B represent the demand for the tracks, the supply of the tracks and the congestion in a bin B, respectively, $\mathcal{N}$ is the set of all the nets in the design, and the summations of the demands and the supplies are carried out over all the bins in the layout. If the track supplies in all the bins are identical, then it is easy to observe that the TWL is proportional to the average congestion.

However, there are several fundamental problems with the use of the TWL as a congestion metric. Different bins often have different supplies in practice, so that their C^B values may not be directly comparable with each other. More significantly, the minimization of the average congestion may not necessarily result in the least congested design possible. This is because the TWL does not not capture the spatial aspects (*i.e.*, the locality) of the congested regions. A design can easily have very low average congestion and yet have a few densely congested bins that may be very difficult to route successfully. For instance, this would be the case if their congestion was largely the result of intra-bin nets that cannot be rerouted to other bins.

Another problem with the use of the TWL at the placement stage is that the actual wirelengths of the nets are not known, since the nets have not yet been routed. Therefore, while computing the TWL, the wirelength of a net is estimated using metrics such as the *half-rectangle perimeter* (HRPM) of its bounding box or the length of the *minimum spanning tree* (MST) for the net. These metrics are oblivious to congestion and do not account for any detours that may subsequently occur while routing the net. Given this large source of inaccuracy, the additional runtime overhead required for the use of better netlength estimates such as rectilinear Steiner trees (RSTs) may often not be justified during the computation of the TWL. Indeed, even MSTs are not used very often during TWL computation. Instead, the TWL computation usually relies on HRPM estimates for the nets.

However, although the HRPM metric is an exact measure for the minimum wirelength of a net that contains two or three pins, it may significantly underestimate even the best possible wirelength required to route a multipin net. This limitation is overcome through the use of empirical compensation factors for the HRPM of a net that depend on the number of pins in the net. This approach, first proposed in the RISA congestion-aware placement engine [Che94], has been widely adopted in many applications that require fast netlength estimation. The work in [Che94] carried out an empirical analysis of the optimal Steiner routes for a large number of randomly generated multipin nets, to measure the average factor by which the wirelength of a Steiner tree for a net with a given number of pins exceeds its HRPM estimate. Table 2.1 presents these compensation multipliers for several different pin counts.

[2] Note that a *net* is a logical concept, whereas a *wire* is its physical implementation obtained after the routing. Following the usual practice in the literature on placement, we use the terms net and wire interchangeably; they can be interpreted appropriately depending on the context.

Pin Count	1–3	4	5	6	8	10	15	20	25	30	35	40	45	50
HRPM Multiplier	1.00	1.08	1.15	1.22	1.34	1.45	1.69	1.89	2.07	2.23	2.39	2.54	2.66	2.79

Table 2.1. Multipliers for the HRPM measure of a net to compensate for multiple pins [Che94].

In spite of the inaccuracies inherent in these fast congestion-oblivious netlength estimates, they exhibit good fidelity with the routed wirelengths of the nets, especially in designs that are not very congested. However, they can be misleading in congested designs in which a significant fraction of the nets are detoured during routing.

Thus, although the use of the TWL as a congestion predictor can involve significant inaccuracies, its advantage is that it involves little computational overhead. This overhead can be reduced further through the judicious application of structural techniques such as those based on Rent's rule, as discussed in Section 2.1.4. However, modern congestion-aware placement engines usually rely on congestion metrics that can provide greater spatial discrimination than that achievable through the scalar TWL metric.

2.1.2 Pin Density

The pin density (also known as the *structural pin density*) metric has been employed by several placement algorithms to improve the routability of congested regions in a design. It is defined for a bin as the ratio of the number of pins in the bin to the area of the bin (or, alternatively, as the number of the pins in the bin if each of the bins has the same area). Thus, for example, the central bin shown in Fig. 2.1 has a pin density of $5/(WH)$ (where W and H are the width and the height of the bin, respectively), since there are five pins in the bin (namely, two belonging to the inverter and three belonging to the two-input NAND gate).

This metric captures the contributions of the intra-bin nets and the inter-bin nets at least one of whose pins lie within the bin. It, however, ignores the flyover wires which are routed through the bin, even though they consume routing resources within the bin. Thus, it models the congestion due to local wires well, but can significantly underestimate the congestion caused due to global wires. However, it has been empirically shown that around 75% of the congestion in a bin is caused due to nets that have a pin in that bin [HM02]. Given the ease of computation of this metric and its fidelity with the actual routing congestion, it is a suitable candidate for congestion estimation during early placement optimizations. Unlike the TWL metric which is a scalar that characterizes the entire design, the pin density metric is quite good at identifying the specific bins that are likely to suffer from congestion.

2.1.3 Perimeter Degree

The perimeter degree of a bin is defined as the ratio of the number of inter-bin nets that have at least one pin inside a bin to the perimeter[3] of the bin. As an example, the central bin shown in Fig. 2.1 has a perimeter degree of $1/(W + H)$, since there are two nets, n_2 and n_4, which have pins inside the bin and are also connected to some cells outside the bin, and the perimeter of the bin is $2(W + H)$.

This metric ignores the routing demand for all intra-bin nets (such as net n_1 in our example), as well as for flyover nets (such as net n_3). Thus, it ignores the congestion due to short, local nets completely and models the global congestion partially. Therefore, it captures the expected congestion at the boundary of the bin rather than that within the bin, in contrast to the pin density metric. Furthermore, compared to the pin density metric, it tends to accentuate the congestion problems in large bins (because the perimeter of a large bin grows less rapidly than its area when compared to a small bin).

As we shall see in Section 2.1.4, this metric lends itself to very efficient approximation through the application of Rent's rule. The perimeter degree has been used for congestion alleviation during placement in [SPK03].

2.1.4 Application of Rent's Rule to Congestion Metrics

The so-called *Rent's rule* [LR71] is an empirical observation about the relationship between the number of terminals required by a design block to interface with its environment and the number of circuit components within the block. This relationship was first observed almost four decades back, and has been shown to hold across a large spectrum of design styles, design sizes, circuit families, and process generations. It can be represented by the following equation:

$$E = AG^r, \tag{2.1}$$

where E is the number of terminals in a block that contains G cells, A is the average number of terminals per cell within the block, and r $(0 \leq r \leq 1)$ is a constant known as the *Rent's exponent.* Although the exact values of A and r may differ from design to design, they appear to hold for any given design across a wide range of block sizes within the design.

Equation (2.1) is found to be valid when the number of partitions of the design is greater than five or so; for fewer partitions, the number of terminals required is smaller than that predicted by Equation (2.1), and is given by a more complex relationship. It has also been found that Rent's rule underestimates the number of interface terminals when the number of cells in

[3] Using a different normalization, [SPK03] defines the perimeter of a bin as the square root of its area, rather than the usual sum of the lengths of all four of its boundaries.

a block is small [Str01]. The domains where Rent's rule over- and underestimates the number of interface terminals are known as *Region II* and *Region III* of the Rent's rule curve, respectively, with *Region I* being used to refer to the domain in which Rent's rule holds. The intuitive explanation for Region II is that designers typically attempt to minimize the number of external pins on a package by time-multiplexing several signals on one pin or by encoding them. In contrast, Region III appears because very small blocks may often be dominated by complex, high fanin cells, or by simple cells such as inverters and buffers that can drive large loads and therefore, have large fanouts.

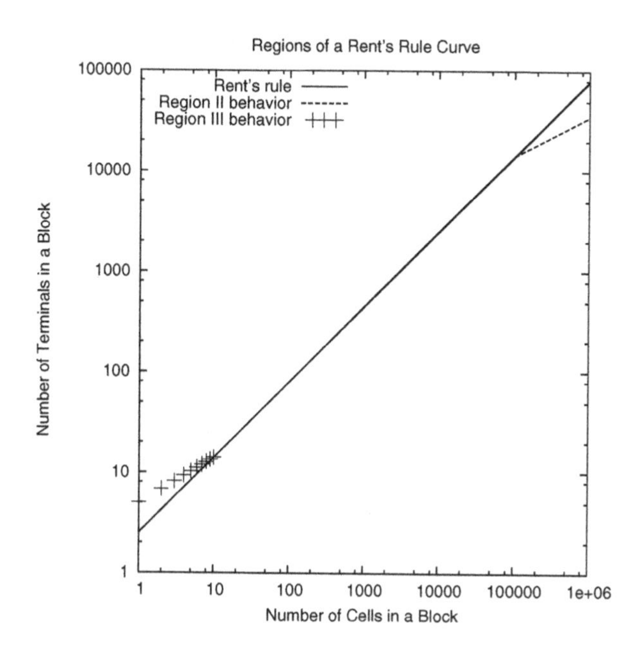

Fig. 2.2. The Rent's rule curve $E = 2.5G^{0.75}$ for a typical design.

An example of such a curve is shown in Fig. 2.2, where the number of cells in partitioned blocks and the number of interface terminals are plotted on the horizontal and vertical axes, respectively, using logarithmic scales. The Rent's rule depicted in the figure represents the equation $E = 2.5G^{0.75}$, where 0.75 is the Rent's exponent and 2.5 is the average number of pins per cell[4]. In this figure, as the number of cells in a block increases along the horizontal axis, one

[4] These values are typical for modern designs; the Rent's exponent for most circuits is greater than 0.50 [Str01], and today's standard cell libraries contain many complex cells with a large number of inputs. One way to obtain the Rent's rule curve for a given design is to partition the design successively to obtain blocks of various sizes, counting the number of interface terminals for each block. The average number of cells and interface terminals for each block size can be plotted

moves from Region III to the main portion of the curve in Region I, and finally to Region II. For this plot, Region II corresponds to the blocks in which the number of cells varies from 10^5 to 10^6, whereas Region III represents blocks containing at most ten cells.

Rent's rule has been used widely for many applications such as *a priori* wirelength estimation[5], placement, and partitioning. Applying Rent's rule, several works such as [Don79, Dav98] have derived the relationship between the total wirelength and Rent's exponent; a good survey of this area of research can be found in [Str01]. The relationship between the average netlength for a partitioning-based placement of a given design in a square area and the Rent's exponent for that design has been shown to be as follows [Don79]:

$$\begin{aligned} \bar{l} &\sim G^{r-\frac{1}{2}}, && r > 1/2, \\ \bar{l} &\sim \log G, && r = 1/2, \\ \bar{l} &\sim f(r), && r < 1/2, \end{aligned}$$

where $\bar{l}$ is the average netlength and $f(r)$ is a function that is independent of G. Most real-world circuits are modeled by the first of the three cases described in these equations. These equations indicate that the larger the Rent's exponent for a design, the higher is the average netlength (and therefore, the TWL) in the design. Since a large value of the TWL metric usually corresponds to increased congestion (as discussed in Section 2.1.1), the Rent's exponent is also an indirect measure of congestion. Indeed, once the Rent's exponent has been precomputed for a design or a family of designs, the approximation of the TWL for any design block of known size requires constant time, with no iteration being required over all the nets in the block.

The direct application of Rent's rule can also be used to speed up the computation of the perimeter degree metric. If the Rent's exponent is known for a given design, then the number of terminals for any cluster of cells can be approximated in constant time. More specifically, given any bin, the number of inter-bin wires that have at least one pin in that bin can be approximated in constant time, without having to iterate over all the cells in the bin. As a result, the perimeter degree metric can be computed for an entire design in $O(B)$ time, where B is the number of bins in the layout.

Of course, the price for the speed-up in the computation of the TWL or the perimeter degree through the application of Rent's rule is an increase in the error inherent in the metric. However, this is usually not a serious concern during the early stages of placement optimization. Indeed, the Rent's exponent has been used as a metric during early placement targeting the TWL or the average congestion in [YKS02].

---

on a log-log plot, followed by curve fitting to obtain the Rent's exponent for the design.

[5] Wirelength estimation schemes that rely solely on the connectivity of the circuit graph and do not use any placement information are referred to as *a priori* schemes.

2.2 Probabilistic Estimation Methods

The only way to obtain a perfectly accurate congestion map is to run the global router on all the placed cells. However, this is an expensive operation. Probabilistic estimation methods (also referred to as *stochastic* methods) have been developed as a fast way to approximate the behavior of global routers. Instead of attempting to find a unique route for each net, probabilistic estimation methods assume that all "reasonable" routes for a net are equally likely, and consider all these routes while computing the congestion contribution of a net to the bins that it may be routed through. Different flavors of probabilistic congestion maps use different notions for what constitutes a reasonable route, although most of them consider only those routes that do not involve any detours.

Since probabilistic estimation techniques avoid choosing between the different routes possible for a given net or even enumerating these routes, they also manage to avoid the combinatorial optimization problem that a global router attempts to solve during the process of routing the nets. In particular, probabilistic estimation is independent of the order in which the nets are considered. As a result, these techniques are considerably faster than the global routing process (since they rely on closed form formulas for their congestion computations). However, this computational efficiency is obtained at the cost of accuracy; real-world global routers can diverge significantly from the simple routing behavior that these techniques model (as shall be discussed in Section 2.2.7). Examples of router behaviors that are not modeled by these techniques include the preferential routing of performance-critical nets, and the ability of routers to avoid congested regions if alternative routes through uncongested regions are available. Yet, in spite of all the inaccuracies in congestion map estimation using probabilistic estimation, these techniques are good candidates for application during the early stages of placement and post-placement circuit optimization. Indeed, they have been used successfully in several commercial physical synthesis tools.

Since routers typically try to route nets using shortest possible paths as far as possible, it is reasonable for probabilistic estimation techniques to ignore routes that involve detours. Furthermore, these techniques avoid the complications of topology generation by decomposing each multipin net into two-pin segments using some simple heuristic model such as a clique, a minimum spanning tree or a rectilinear Steiner tree. In the same vein, the layer assignment of the routes is also ignored. For any given net, a probabilistic estimation technique considers all the valid routes for that net that satisfy the modeling assumptions for that technique. The congestion contribution of each such route to every bin that it passes through is then weighted by the probability of that route being selected, using some simple probability distribution such as a uniform distribution.

Although several probabilistic estimation models have been explored, the two that have received the most attention are distinguished by the number of bends that they allow in their routings. The more general model of the two permits an arbitrary number of bends in the routes [LTK+02], in contrast to the other model that considers only those routes that have at most two bends [CZY+99] [WBG04]. Routes that involve just a single bend are said to be *L-shaped*, whereas those that involve two bends are said to be *Z-shaped*, as shown in Fig. 2.3. Routes whose source and sink pins lie in the same row or column of bins are said to be *flat*. Given a choice of two routes having the same wirelength but different numbers of vias, most routers will select the one with the fewer vias (and consequently, fewer bends). Therefore, the probabilistic estimation model that restricts its routes to those with at most two bends usually does a better job of modeling actual router behavior. We will next discuss a probabilistic estimation technique that uses such a model, and subsequently briefly explore the more general multibend model in Section 2.2.4.

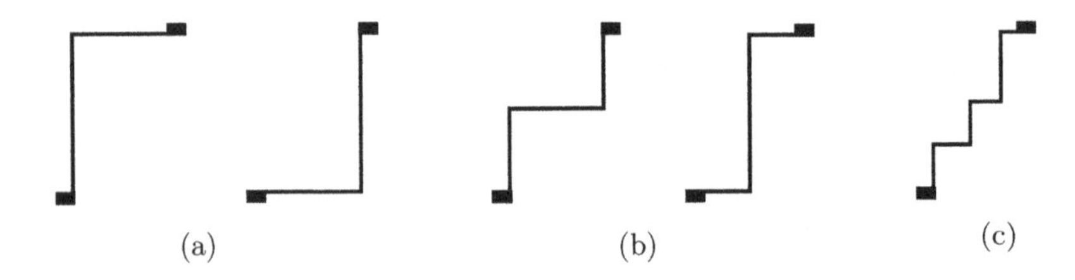

Fig. 2.3. Examples of (a) L-shaped, (b) Z-shaped, and (c) multibend routes.

Given a two-pin net and a bin, the general procedure for probabilistic congestion map generation attempts to obtain an expression for the expected routing demand (also referred to as the *utilization*) of the net in that bin. This is achieved by computing the fraction of valid routes for the net that pass through the bin. More precisely, we weight the track usage within the bin for some given route by the probability of that route being selected. All these weighted track usages are then summed up over all the routes of the net to obtain the routing demand of the net in the bin.

We illustrate this computation first for intra-bin nets and flat nets, and then extend it to the case of general L-shaped and Z-shaped nets. Let the bins created by the tessellation of the layout area be indexed by their column and row indices, with the bin $(1, 1)$ lying in the lower left corner of the layout. Without loss of generality, we assume that the net whose contribution to the congestion map is being computed has its pins in the bins $(1, 1)$ and (m, k). Let the utilization of the net in some bin (i, j) (which lies in the ith column and the jth row) be denoted by $U^{(i,j)}$ (with $U_x^{(i,j)}$ and $U_y^{(i,j)}$ referring to routing demands in the horizontal and vertical directions, respectively). For the sake of simplicity, we assume that all bins have the same width and height, denoted by W and H, respectively; the extension to bins of different sizes is straightforward. Furthermore, we assume that the two pins of a net are denoted by a and b, and that these pins lie at locations whose coordinates are (x_a, y_a) and (x_b, y_b), respectively. Let the horizontal (vertical) distance of pin a from the right (upper) boundary of the bin $(1, 1)$ that contains it be denoted by d_x^a (d_y^a), and the horizontal (vertical) distance of pin b from the left (lower) boundary of the bin (m, k) that contains it be denoted by d_x^b (d_y^b), as illustrated in Figs. 2.5 and 2.6.

2.2.1 Intra-bin Nets

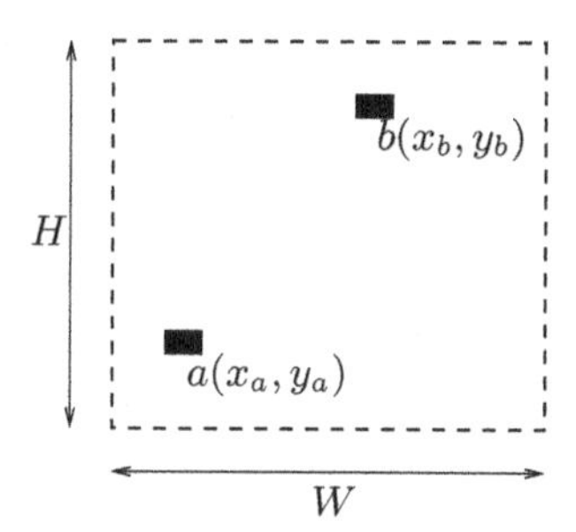

Fig. 2.4. An intra-bin net with pins at (x_a, y_a) and (x_b, y_b).

The shortest possible route for the intra-bin net in Fig. 2.4 requires a horizontal track of length $|x_a - x_b|$ and a vertical track of length $|y_a - y_b|$. Therefore, the horizontal and vertical routing demands in the bin due to the net are respectively given by:

$$U_x^{(1,1)} = \frac{|x_a - x_b|}{W} \text{ and } U_y^{(1,1)} = \frac{|y_a - y_b|}{H}.$$

2.2.2 Flat Nets

As mentioned earlier, flat nets are inter-bin nets whose pins lie either in the same row or in the same column of bins. A flat net is said to be horizontally flat if its terminals lie in the same row; otherwise, it is said to be vertically flat. Horizontally and vertically flat nets are illustrated in Fig. 2.5(a) and (b), respectively. We assume that these types of nets are routed with at most one bend. This can be accomplished using either of two L-shaped paths, shown by the solid and dashed-dotted lines in the figure. Therefore, for a horizontally (vertically) flat net, vertical (horizontal) demand may exist only in the first and last bins.

As can be seen in Fig. 2.5(a), the horizontal track length required in the first bin, namely $(1, 1)$, equals the distance of the pin a from the right boundary of the bin (*i.e.*, d_x^a). Similarly, the horizontal track length required in the

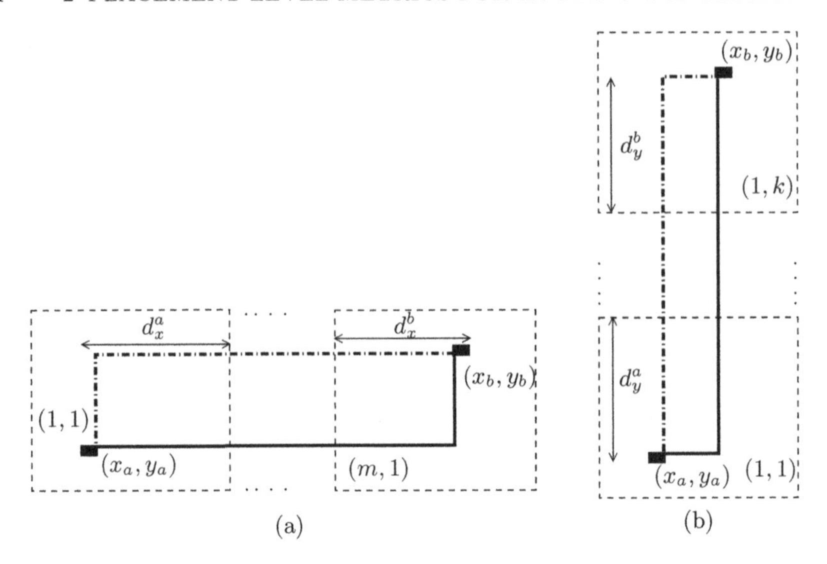

Fig. 2.5. Routing demand analysis for (a) horizontally flat and (b) vertically flat nets.

last bin, $(m, 1)$, is d_x^b, as shown in the figure. The net requires an entire horizontal track in each of the middle bins that it passes through. Therefore, the horizontal routing demand for this horizontally flat net is given by:

$$
\begin{aligned}
U_x^{(1,1)} &= \frac{d_x^a}{W}, \\
U_x^{(j,1)} &= 1, \text{ for } j = 2, \ldots, m-1, \text{ and}, \\
U_x^{(m,1)} &= \frac{d_x^b}{W},
\end{aligned}
$$

where W is the width of a bin. For this net, vertical routing track demand exists only in the first or last bins, each with probability 1/2, and is given by:

$$
\begin{aligned}
U_y^{(1,1)} &= \frac{|y_a - y_b|}{2H}, \\
U_y^{(j,1)} &= 0, \text{ for } j = 2, \ldots, m-1, \text{ and}, \\
U_y^{(m,1)} &= \frac{|y_a - y_b|}{2H},
\end{aligned}
$$

where H is the height of a bin.

Similarly, the horizontal and vertical routing demands for the vertically flat net in Fig. 2.5(b) can be written as:

$$
\begin{aligned}
U_x^{(1,1)} &= \frac{|x_a - x_b|}{2W}, \\
U_x^{(1,j)} &= 0, \text{ for } j = 2, \ldots, k-1, \\
U_x^{(1,k)} &= \frac{|x_a - x_b|}{2W}, \\
U_y^{(1,1)} &= \frac{d_y^a}{H}, \\
U_y^{(1,j)} &= 1, \text{ for } j = 2, \ldots, k-1, \text{ and,} \\
U_y^{(1,k)} &= \frac{d_y^b}{H}.
\end{aligned}
$$

2.2.3 Single and Double Bend Routes for Inter-bin Nets

In the routing model assumed in [WBG04], inter-bin routes that are not flat are assumed to be routed with at most two bends, forming L-shaped or Z-shaped routes. Let $U^{p,(i,j)}$ represent the routing demand in bin (i,j) obtained by considering only p-bend routes for a net. Then, the overall utilization $U^{(i,j)}$ in bin (i,j) for the net can be written as:

$$U^{(i,j)} = \alpha_1 U^{1,(i,j)} + \alpha_2 U^{2,(i,j)},$$

where α_1 and α_2 are empirically chosen weights indicating the relative preferences for single and double bend routes, respectively. Typically, $\alpha_1 \geq \alpha_2$ (because, given a choice, routers prefer routes with fewer vias), $\alpha_1 + \alpha_2 = 1$, and $\alpha_1, \alpha_2 \geq 0$ (in order to allow the interpretation of these weights as probabilities).

For a net with pins $a(x_a, y_a)$ in bin $(1,1)$ and $b(x_b, y_b)$ in bin (m,k), there are two possible single bend routes, whereas the number of double bend routes is $(m + k - 4)$ (assuming $m, k > 1$). These routes lead to different routing demands in different bins lying within the bounding box of the net. The computation of the routing demand in all these bins can be covered by the analysis of nine different cases, based on the location of the bin relative to the pins of the net. These cases include the four bins located at the corners of the bounding box, bins located along the four sides of the bounding box but not at its corners, and the bins located in the interior of the bounding box.

Let us first consider the bin $(1,1)$, located at the lower left corner of the bounding box of the net. In this case, all the routes to the destination bin (m,k) leave either horizontally or vertically from this bin. The numbers of routes leaving this bin horizontally and vertically are $(m-1)$ and $(k-1)$, respectively, as shown in Fig. 2.6(a). Of these routes, one route in either direction is L-shaped; these two single bend routes pass through the bins lying along the edge of the bounding box of the net. The remaining $(m + k - 4)$ routes are all Z-shaped. As was discussed in Section 2.2.2, we can see that the routes leaving the bin horizontally require a horizontal track of length d_x^a,

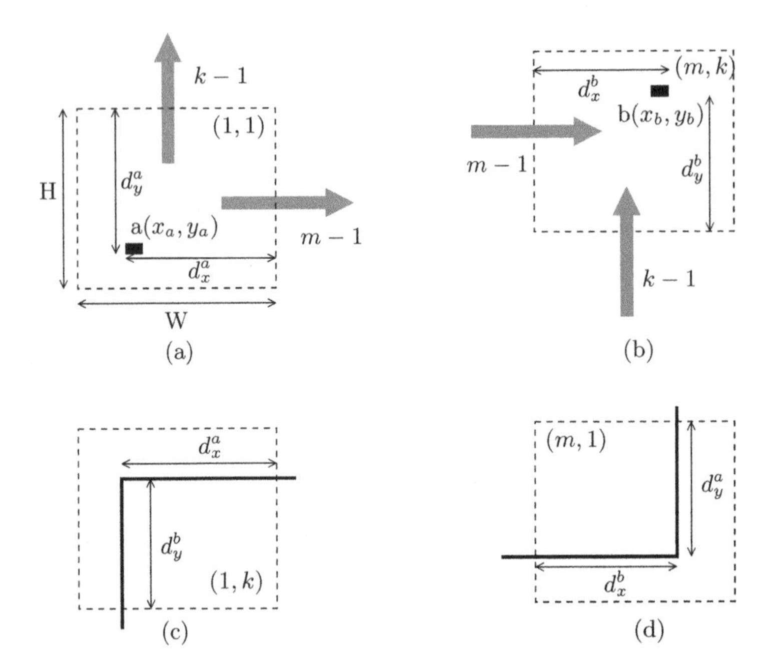

Fig. 2.6. Routing demand analysis for bins located at the corners of the bounding box of a net routed from bin $(1, 1)$ to bin (m, k): (a) Bin $(1, 1)$ at lower left corner, (b) Bin (m, k) at upper right corner, (c) Bin $(1, k)$ at upper left corner, and, (d) Bin $(m, 1)$ at lower right corner.

whereas routes departing vertically use a vertical track of length d_y^a. Therefore, the contribution to the routing utilization due to the L-shaped paths is given by:

$$U_x^{1,(1,1)} = \frac{d_x^a}{2W} \text{ and } U_y^{1,(1,1)} = \frac{d_y^a}{2H}.$$

Similarly, the contribution due to Z-shaped paths is given by:

$$U_x^{2,(1,1)} = \frac{m-2}{m+k-4} \times \frac{d_x^a}{W}, \text{ and,}$$
$$U_y^{2,(1,1)} = \frac{k-2}{m+k-4} \times \frac{d_y^a}{H}.$$

These expressions can be combined to yield the overall routing demand in bin $(1, 1)$ as follows:

$$U_x^{(1,1)} = \alpha_1 U_x^{1,(1,1)} + \alpha_2 U_x^{2,(1,1)}$$
$$U_y^{(1,1)} = \alpha_1 U_y^{1,(1,1)} + \alpha_2 U_y^{2,(1,1)}.$$

The analysis of the routing demand for bin (m, k), located in the top right corner of the bounding box and illustrated in Fig. 2.6(b), is analogous to that for bin $(1, 1)$ discussed above.

The only routes that pass through the bins located in the upper left and lower right corners of the bounding box (illustrated in Fig. 2.6(c) and (d)) are the single bend routes. Therefore, the utilization for the bin in the upper left corner can easily be shown to be:

$$U_x^{(1,k)} = \alpha_1 \frac{d_x^a}{2W} \text{ and } U_y^{(1,k)} = \alpha_1 \frac{d_y^b}{2H}.$$

Similarly, the utilization for bin $(m, 1)$, in the lower right corner, can be derived as:

$$U_x^{(m,1)} = \alpha_1 \frac{d_x^b}{2W} \text{ and } U_y^{(m,1)} = \alpha_1 \frac{d_y^a}{2H}.$$

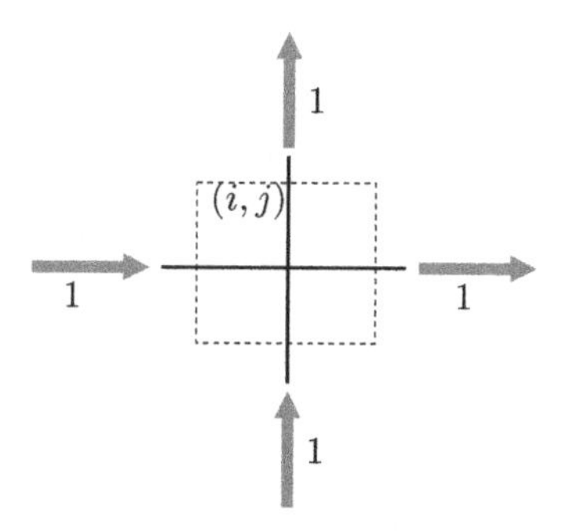

Fig. 2.7. Routing demand analysis for a bin located within the interior of the bounding box of a net.

Next, let us analyze the utilization in a bin (i, j) (with $1 < i < m$ and $1 < j < k$) that lies in the interior of the bounding box of the net. As shown in Fig. 2.7, one Z-shaped route enters the bin horizontally and one enters it vertically; these two routes leave without any bends, using up one horizontal and one vertical track in the process. Therefore, the horizontal and vertical routing demand in the bin is given by:

$$U_x^{(i,j)} = \alpha_2 \frac{1}{m+k-4} \text{ and } U_y^{(i,j)} = \alpha_2 \frac{1}{m+k-4}.$$

Now, consider the non-corner bins located in the leftmost column of the bounding box of the net. In other words, let us derive the utilization for a bin $(1, j)$ with $1 < j < k$. One of the two L-shaped routes passes through this bin, entering and exiting vertically. Of the $(k-2)$ Z-shaped routes whose middle segments are horizontal, $(k-j)$ routes enter this bin, across its lower boundary. One of these Z-shaped routes turns right and exits the bin horizontally, whereas the remaining $(k-j-1)$ routes continue vertically (to turn right at some bin $(1, j')$ with $j < j' < k$). The Z-shaped route that enters the bin vertically and leaves it horizontally requires a horizontal track of length d_x^a and half of a vertical track. The remaining routes passing through this bin use up one full vertical track each. Therefore, the horizontal routing demand is given by:

$$U_x^{(1,j)} = \alpha_2 \frac{d_x^a}{(m+k-4)W}.$$

The vertical routing demand due to the L-shaped route passing through the bin is given by:

$$U_y^{1,(1,j)} = \frac{1}{2},$$

whereas that due to the Z-shaped routes is given by:

$$U_y^{2,(1,j)} = \frac{1}{2(m+k-4)} + \frac{k-j-1}{m+k-4}.$$

Therefore, the total vertical routing utilization in the bin is given by:

$$\begin{aligned} U_y^{(1,j)} &= \alpha_1 U_y^{1,(1,j)} + \alpha_2 U_y^{2,(1,j)} \\ &= \frac{1}{2}\alpha_1 + \alpha_2 \frac{2(k-j-1)+1}{2(m+k-4)}. \end{aligned}$$

The analysis of the utilization in the non-corner bins located along the remaining three edges of the bounding box of the net is analogous to the case discussed above, and is omitted for brevity.

2.2.4 Multibend Routes for Inter-bin Nets

The method discussed in the previous section for single and double bend routes can also be extended to consider all minimum length multibend routes. For each bin within the bounding box of a net, the routing demand can be derived by counting the multibend routes passing through that bin, as described in [LTK+02]. The number of shortest possible multibend routes from bin $(1,1)$ to bin (m,k) is $\binom{m+k-2}{m-1}$, in contrast to $(m+k-2)$ routes under the single and double bend model. The demand due to multibend routes can be weighed uniformly as in [LTK+02], or based on the number of bends as discussed in Section 2.2.3.

Since the routing model that considers multibend routes explores a larger space than one that considers single and double bend routes only, it leads to a different distribution of the routing demands. Typically, for an inter-bin net, estimates based on single and double bend routes predict high utilization in bins lying along the edges of the bounding box of the net, whereas those based on multibend routes show increased routing demand in the interior of the bounding box. Moreover, the bins that are in the interior of the bounding box have the uniform routing demand $\alpha_2/(m+k-4)$ under the single and

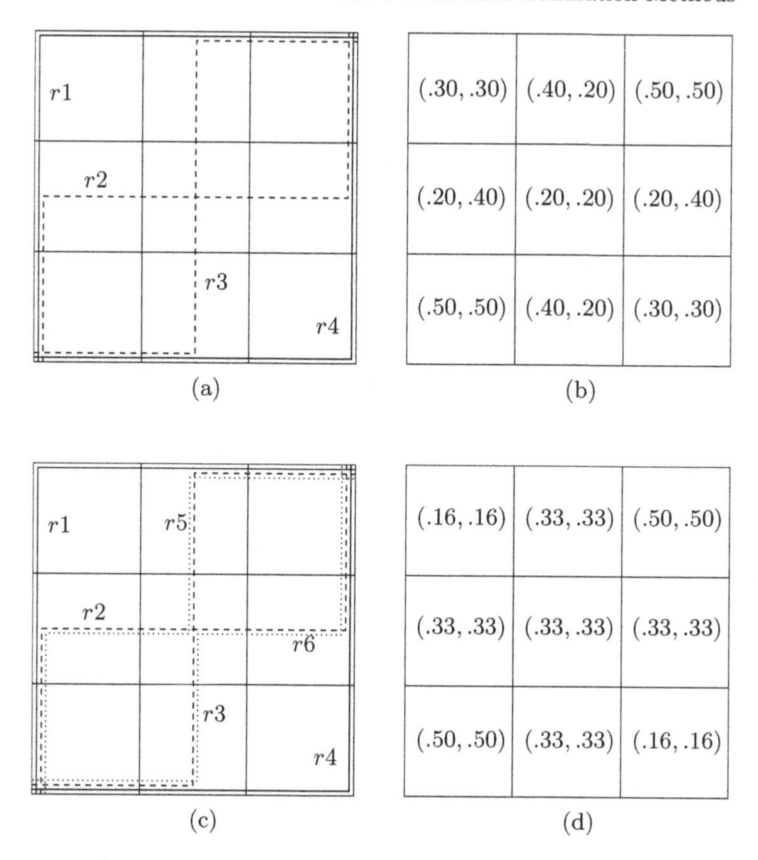

Fig. 2.8. (a) L- and Z-shaped routes, and (b) the corresponding routing demands. (c) Multibend routes, and (d) the corresponding routing demands.

double bend route model, whereas the consideration of multibend routes leads to varying routing demands in these bins.

As an example, Fig. 2.8 illustrates the differences in the distributions of the routing demands on a 3×3 grid obtained using probabilistic estimation methods based on single and double bend routes only (in Fig. 2.8 (a) and (b)) and multibend routes (in Fig. 2.8(c) and (d)). In this example, the values of α_1 and α_2 for the former estimation model are taken to be 0.6 and 0.4, respectively, whereas the multibend model uses uniform weighting for all the routes. In the figure, single bend routes (namely, $r1$ and $r4$) are depicted using solid lines, double bend routes (namely, $r2$ and $r3$) are shown using dashed lines, whereas dotted lines are used for the remaining two routes (namely, $r5$ and $r6$) that involve three bends each. The tuple (U_x, U_y) within a bin represents the horizonal and vertical routing demand within that bin. It can be seen that the single and double bend model yields the least routing demand in the central bin (since only two routes pass through that bin, and these routes are non-preferred). In contrast, the least routing demand with the multibend model shows up in the two non-pin corners of the bounding box. Thus, the two routing models lead to different probabilistic utilizations, and therefore different predictions about the locations of congested hot spots.

In layouts that are not very congested, the tendency of routers to minimize the usage of vias makes the congestion maps predicted using only single and double bend routes more representative of the actual post-routing congestion map. On the other hand, although nets in congested regions are often routed with more bends, the response of the router to existing congestion and the increased occurrence of detours in such regions increases the error inherent in probabilistic congestion map prediction irrespective of the routing model used.

2.2.5 Routing Blockage Models

The modeling of routing blockages is one of the most challenging issues faced by all probabilistic estimation methods. A routing blockage is an area where the routing resources are either reduced or unavailable. Modern designs typically include many blockages due to the presence of custom macros or hard[6] intellectual property (IP) blocks. If no routing tracks are available across the blockage, then the blockage is said to be complete; otherwise, it is partial. Partial blockages occur frequently due to prerouted signal and clock nets and the power grid, and can be modeled with relative ease by reducing the routing supplies in the corresponding bins. The modeling of complete blockages is much harder, and relies on various heuristics to mimic the behavior of a typical router in the vicinity of such blockages. For the sake of convenience, we refer to complete blockages merely as blockages in the remainder of this chapter (since partial blockages do not require any special handling).

Since a router cannot use any tracks in a blocked bin, any probabilistic utilization within such a bin should also be zero. For nets some of whose minimum length routes pass through blocked bins, it is reasonable to assume that a router would try to find a minimum length route through neighboring unblocked bins if such a path exists. Therefore, for a net whose bounding box includes one or more blocked bins but that also has some minimum length route passing only through unblocked bins, the routing demand in the blocked bins is distributed to their neighboring bins to reflect the expected behavior of the router.

Consider the example depicted in Fig. 2.9(a), in which the central bin is blocked, the blockage being represented by the hexagonal pattern. If this blockage is ignored, the routing demand (assuming a single and double bend

---

[6] IP blocks available only in layout formats such as GDSII that are not amenable to any changes are said to be *hard*. In contrast, *soft* IP blocks are available at a higher abstraction level, such as that of a register transfer level (RTL) description, and designers have the freedom to implement or modify their layout.

routing model) is as shown in Fig. 2.9(b) (which is the same as Fig. 2.8(b)). The routing demand for the blocked bin can be distributed among its unblocked neighbors using some heuristic that weights this redistribution by the distance between the bins [LTK+02,SY05]. For instance, it may be distributed equally among the four bins that lie within one unit of Manhattan distance, as shown in Fig. 2.9(c). However, it is easy to see that the only minimum length routes possible in this case are the two L-shaped routes; the corresponding routing demand is depicted in Fig. 2.9(d). Unfortunately, the price for the increased accuracy obtained after determining that only the two L-shaped routes are possible is an increase in computation time.

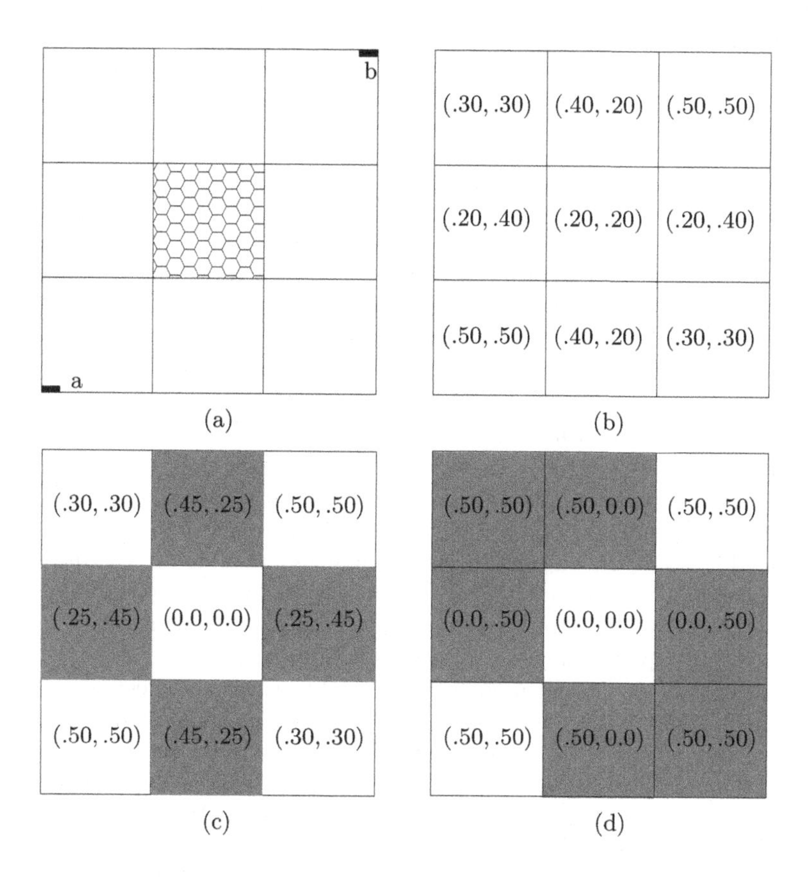

Fig. 2.9. (a) An example of a blocked bin. (b) Routing demand computed ignoring the blockage. (c) Routing demand after heuristically distributing the routing demand from the blocked bin to adjacent unblocked ones. (d) A more realistic routing utilization.

However, if no minimum length route for a net can avoid blocked bins, the router can be expected to try to complete the routing of the net with the shortest possible detour. Depending on the complexity of the blockage, the detour can be modeled during probabilistic congestion map estimation either by creating pseudo-pins on the net or by performing explicit routing.

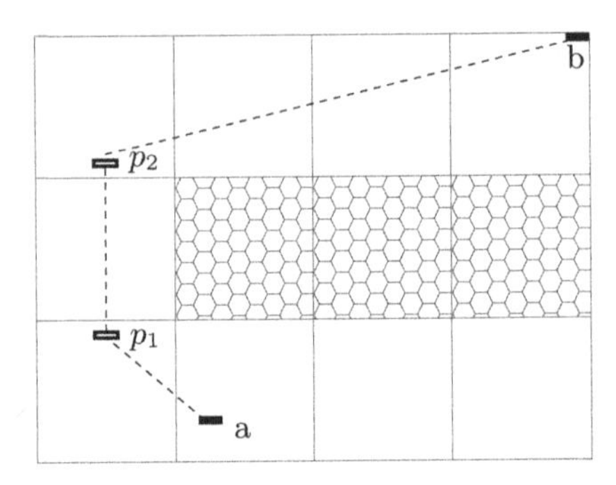

Fig. 2.10. Modeling a detour due to a row blockage using pseudo-pins.

An example of such a detour is depicted in Fig. 2.10. In this example, one row inside the bounding box is completely blocked, eliminating all minimum length routes for the net. Such a blockage of an entire row (column) within the bounding box can be handled by expanding the bounding box in the horizontal (vertical) direction until an unblocked bin is found in the blocked row (column), and then inserting pseudo-pins on the net to route it through this unblocked bin. Figure 2.10 shows pseudo-pins p_1 and p_2 created after expanding the bounding box to the left; as a result, the net is decomposed into three distinct segments. The estimation of probabilistic routing utilization is performed individually on each of these segments and added to the respective bins. This heuristic typically has a faster runtime than maze routing. However, it may not be applicable to more complicated blockage configurations, leaving no option but explicit routing to handle such blockages.

In general, heuristics for modeling routing blockages are effective at allowing reasonably accurate congestion estimates without excessive computation overhead only when the blockages are simple and few. In the presence of a large number of complicated blockages, however, probabilistic estimation methods are highly inaccurate because of the modeling of the blockages; a (fast) global router is the only known reliable alternative in such a case.

2.2.6 Complexity of Probabilistic Methods

The pseudocode for a generic probabilistic congestion estimation procedure is shown in Algorithm 1. It begins with the computation of the supplies of routing tracks for all the bins in the layout. This step requires $O(b)$ time, where b is the number of bins in the layout. This is followed by the estimation

Algorithm 1 Congestion estimation using probabilistic methods

```
 1: for all bins B in the layout do
 2:    Compute the routing track supply in B
 3: end for
 4: for all nets N in the design do
 5:    if there are blocked bins in the bounding box of N then
 6:       Apply blockage modeling heuristics to compute utilization for N
 7:    else
 8:       Decompose N into two-pin segments
 9:       for all segments s of N do
10:          for all bins B in the bounding box of s do
11:             Compute probabilistic routing utilization for s in B
12:             Add utilization for s in B to the total utilization of B
13:          end for
14:       end for
15:    end if
16: end for
17: for all bins B in the layout do
18:    Divide the probabilistic routing utilization of B by its available track supply
19: end for
```

of routing demand due to all the nets in the design. Nets involving blocked bins are handled heuristically. All nets are decomposed into two-pin segments, and the probabilistic demand is computed for each of these segments in each of the bins within their bounding boxes.

The decomposition of multipin nets into two-pin segments can be achieved by constructing a minimum spanning tree (MST) or rectilinear Steiner tree (RST). Although RST construction is closer to the actual topology generation used during routing, the runtime overhead for RSTs cannot always justify the additional accuracy, if any, given the errors already inherent in the probabilistic estimation process. Furthermore, the use of the RST may even worsen the accuracy if the RST topologies assumed during congestion estimation and constructed during routing are very different. Therefore, the use of a MST algorithm usually suffices for the decomposition of a net. A MST for a net with p pins can be constructed in $O(p^2)$ time using Prim's algorithm implemented with a Fibonacci heap.

The computation and addition of the probabilistic routing demand for a two-pin net requires $O(b)$ time, since the bounding box of a net may span the entire layout area in the worst case. Therefore, assuming MST construction and a fast heuristic to handle blockages, the **for** loop in lines 4–16 of the pseudocode requires $O(n(bp+p^2))$ time, where n is the number of nets in the design. The subsequent division of the utilizations by the track supplies to compute the expected congestion in each bin requires $O(b)$ time. Therefore, if the maximum number of pins in a net is assumed to be a constant (as is usually true in practice because of fanout constraints during circuit optimization), the overall complexity of probabilistic congestion estimation is $O(nb)$. Thus, in the presence of a few, relatively simple blockages, this complexity is linear in the number of nets. However, if a large number of complicated blockages are present, the overall time complexity for probabilistic estimation may trend towards that of global routing, since many nets may require routing to compute the utilization.

2.2.7 Approximations Inherent in Probabilistic Methods

The price that probabilistic methods pay for efficiency when compared to routing-based congestion estimation methods is an inability to capture the behavior of routers on nets that are difficult to route. This includes approximations in the handling of blockages, limited or non-existent modeling of detours, layer assignment and blockages due to via stacks, as well as approximations in the topology generation for multipin nets. Another significant source of error is the failure of these schemes to model the response of a router to existing congestion, as discussed next.

Routers typically are congestion-aware. In other words, when a router finds that the routing demands in certain bins in the bounding box of a net are approaching or exceeding the available supply, it avoids such bins and selects uncongested bins as much as possible. In contrast, most probabilistic congestion estimation schemes ignore any prior congestion, adding the routing demand for a new net to already congested bins also, even though a route may exist through uncongested bins within the bounding box of the net. This behavior is illustrated by an example in Fig. 2.11. Assume that the horizontal and vertical supply in each bin consists of five tracks each. In this example, Fig. 2.11(a) depicts the prior congestion in each bin within the bounding box of the net. In such a scenario, most routers will route the net through the uncongested bins as shown in Fig. 2.11(b). However, probabilistic routing demand estimation techniques distribute the routing demand for this net among all the bins, even if they are congested; the congestion map obtained from such a scheme using a single and double bend model is shown in Fig. 2.11(c).

This example illustrates how the probabilistic estimates can be pessimistic in densely congested regions. This pessimism in probabilistic congestion maps has the following implications on design convergence strategies:

- If the probabilistic estimation reports no congested areas, then a detour-free global routing solution definitely exists, and one can proceed to the routing stage with the existing placement rather than attempting to improve the congestion further at the placement stage.
- If the probabilistic methods report congested regions, then a detour-free routing solution may still be possible, since the methods overestimate the routing demands in the congested regions. One may proceed to the routing stage if the reported congestion is not too high or if there are only a few, small congested regions.

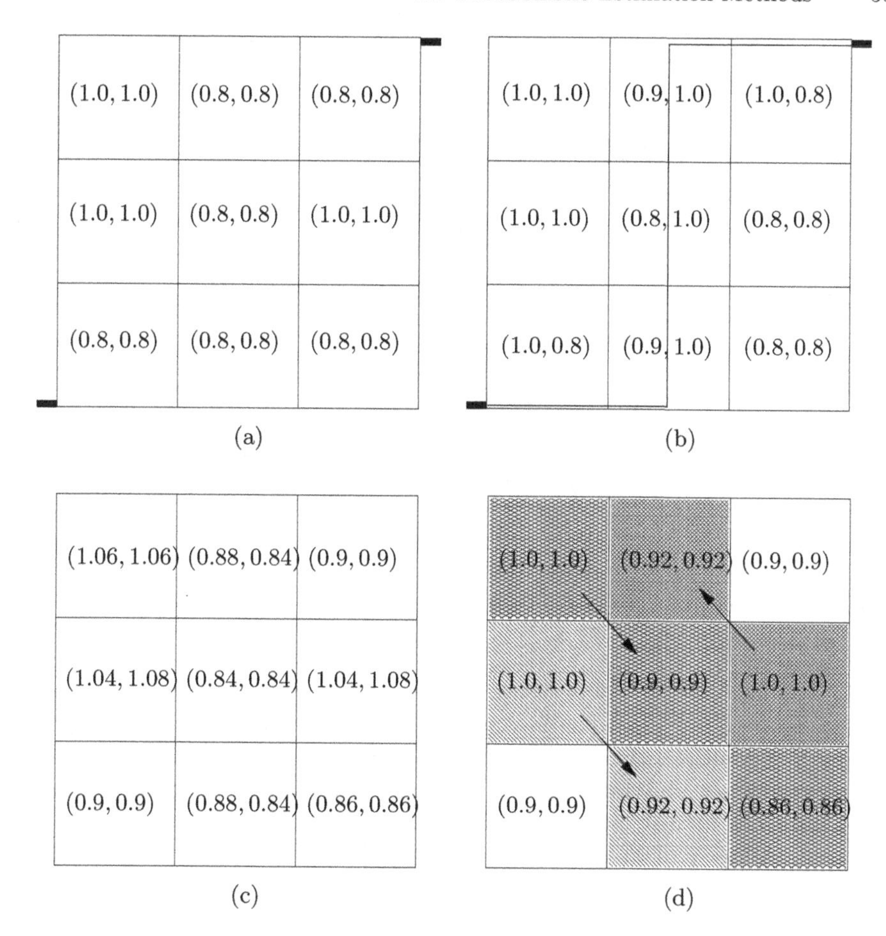

Fig. 2.11. (a) The existing congestion map within the bounding box of a net, before this net is processed. (b) A routing for the net that passes through uncongested bins, and the updated congestion map. (c) Congestion map predicted by a probabilistic congestion estimation scheme after processing this net. (d) Congestion map obtained after congestion redistribution.

The pessimism in the probabilistic congestion maps can be reduced by applying post-processing techniques that redistribute the routing demand from the densely congested bins to the sparsely congested ones [KX03, SY05]. Figure 2.11(d) shows the effect of such a congestion redistribution heuristic proposed in [SY05]. This heuristic buckets the bins within the bounding box of a net into partitions obtained through a breadth-first traversal from one of the pins, such that every minimum length route for the net will pass through exactly one bin in each partition. It then redistributes the routing demand individually within each partition, moving it greedily from overcongested bins to sparsely congested ones. The resulting congestion map is a better approximation of the actual post-routing congestion map (such as the one in Fig. 2.11(b)) than the original map. However, any post-processing heuristic that improves the accuracy and lessens the pessimism in probabilistic congestion estimates comes at the cost of an increase in runtime.

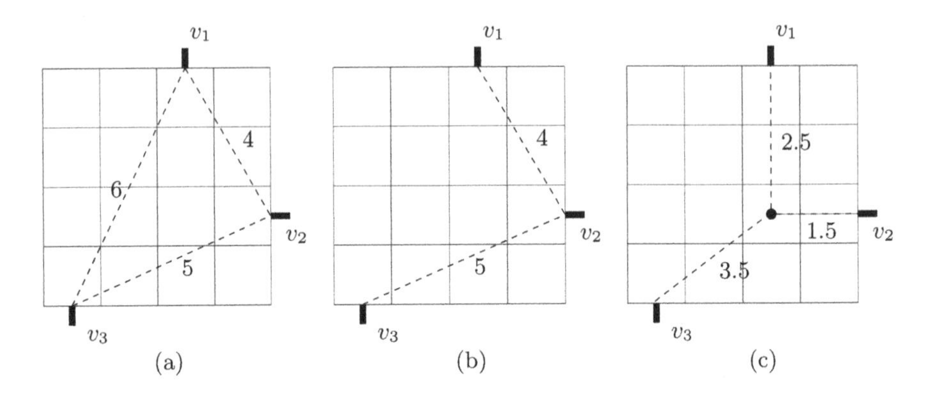

Fig. 2.12. (a) A three-terminal net with a clique on it. (b) An MST with length of 9. (c) An RST with length of 7.5.

The inaccuracies in a probabilistic congestion map that arise because of the modeling of multipin nets can be illustrated through an example. Consider the three-pin net illustrated in Fig. 2.12; the figure depicts three possible decompositions for this net. Assume for the sake of simplicity that each bin is a square whose side is of unit length. It can be seen that the clique model yields the largest netlength (of fifteen units), whereas the RST decomposition results in the shortest netlength, of 7.5 units. The netlength of the MST model of the net is nine units. In general, a congestion map obtained using MST decompositions will point to more congestion in the layout than one that relies on RST decompositions.

2.3 Estimation based on Fast Global Routing

As we observed in Section 2.2.7, probabilistic congestion estimation suffers from several significant sources of errors. For instance, the only viable alternative to deal with complicated blockages is to carry out routing in their vicinity. It is natural to investigate whether the more extensive use of routing can help improve the accuracy of the predicted congestion maps. Of course, it is not practical to invoke a full-fledged global router inside a placement optimization loop to alleviate the congestion, because of large runtime overheads. However, if the global routing can be carried out in a "low effort" mode, it may help generate a congestion map prediction that is more accurate than one obtained using probabilistic estimation. This has motivated the recent development of "fast" global routing techniques targeted primarily towards congestion estimation.

Typically, in modern process technologies, which provide up to nine metal layers at the 90 nm technology node, global routing and layer assignment is performed simultaneously, even though performance-driven global routing in the presence of blockages and performance-driven layer assignment are individually intractable problems [HS01,SL01]. As was discussed in Section 1.1.1, the simultaneous exploration of the layer and bin spaces results in better routes than those obtained by searching these spaces sequentially. However, the price for this simultaneous exploration is paid in high runtimes. Routing graphs in modern routing architectures have an order of magnitude more nodes and edges due to the increased number of metal layers as compared to those from a couple of decades ago. Most global routers still use some version of maze routing or Dijkstra's shortest path algorithm, although often augmented with fast search algorithms, to find routes having the least cost (which may be delay, congestion, wirelength, or some combination of these). When global routing is employed for congestion estimation purposes, some inaccuracy in the predicted routes for the nets (as compared to their actual routes) can be tolerated, especially if it leads to a significant gain in computation time. Therefore, efforts to apply global routing to the problem of congestion estimation have focused primarily on two strategies, namely:

- the reduction of the search space through a coarsening of the routing graph, and,
- the extensive use of fast search algorithms.

2.3.1 Search Space Reduction

The availability of a growing number of routing layers causes the routing graph to be large. Its size can be reduced significantly by collapsing all the horizontal layers and all the vertical layers into two orthogonal layers. This results in a significant reduction in the size of the routing graph. The horizontal (vertical) track supplies for a bin in this collapsed routing graph are obtained by adding the respective contributions due to each horizontal (vertical) layer.

Of course, routes on the collapsed routing graph require far fewer vias than they would in the full multilayer routing graph. Therefore, congestion predictions based on the collapsed routing graph underestimate the effect of blockages created due to via stacks. Another source of inaccuracy in these predictions is the effect of layer assignment on the delays of nets in performance-driven routing. Since the electrical characteristics of different layers can differ significantly in modern process technologies, lumping them together in the collapsed routing graph can cause the router behavior to differ significantly from that on the original routing graph.

Another technique for reducing the size of the routing graph is to impose a coarser tessellation on the layout area than the one used to generate the bins for the actual global routing. Although this reduces the spatial resolution of the congestion measurements, this loss of resolution may be a small price to pay for a significant speedup in the global routing when used for congestion estimation.

2.3.2 Fast Search Algorithms

Given a routing graph, the routing of a net involves the generation of a topology for it, and the embedding of each of the two-pin segments in that topology into the routing graph. The congestion estimation mode can use much simpler and faster topology generation algorithms than those used during the actual global routing process, even if it results in topologies whose embeddings into the routing graph have poorer wirelengths and critical sink delays. Furthermore, even while routing the two-pin segments of the topologies, there are at least three basic techniques that have been used to speed up the global routing process for the purpose of congestion estimation. They are:

- a significantly reduced application of rip-up and reroute heuristics,
- the use of fast routers that do not guarantee shortest routes, and,
- the use of fast search algorithms that guarantee shortest routes.

As will be discussed in Section 4.1 in Chapter 4, most industrial global routers rely heavily on finely tuned rip-up and reroute heuristics for route completion. However, the repeated rip-up and rerouting of a net can add significantly to the runtime of the router. When used for congestion estimation, these heuristics are used much more sparingly in a "low effort" mode of the router. As a consequence, although the runtime of the router is improved significantly, the quality of the routing, as measured by the minimization of routing overflows in the bins, degrades significantly, becoming much more dependent on the order in which the nets have been routed. Therefore, the predicted route for a net as obtained during the congestion estimation mode can be quite different from its actual route generated during the routing stage. This can lead to inaccuracies in the predicted congestion map, with routable designs being identified as congested because of insufficient exploration of the search space through rip-up and reroute strategies.

Unlike global routing that relies heavily on search algorithms that guarantee shortest paths, one can also use the faster but often suboptimal line probe search [Hig69] and its variants to route the nets during the congestion estimation mode. Although these algorithms are much faster than the usual breadth-first search used for routing and are often close to optimal in sparsely congested regions that have few or no blockages, they can perform very poorly in congested regions or regions that are fragmented by numerous complicated blockages. In such regions, they may fail to find a route even if it exists, or find one with a length that is much larger than the optimal length for that routing. Therefore, in scenarios where a line probe search has failed or found a route that is much longer than the bounding box of the net, it may be useful to fall back upon a standard search algorithm that guarantees shortest routes. As long as this falling back does not occur frequently in practice, the use of a line probe router can offer significant speedups to the operation of the global router when used for congestion estimation, with tolerable deterioration in routing quality (even though this deterioration may be unacceptable for the actual global routing process). Of course, as with the reduced application of rip-up and reroute heuristics, the use of fast but non-optimal routers may cause the predicted route of a net to differ from its actual route.

The standard optimal shortest path algorithm used in global routers is based on Dijkstra's algorithm [Dij59] and has $O(|E| + |V| \log |V|)$ time complexity[7] (when implemented using Fibonacci heaps), where $|E|$ and $|V|$ are the number of edges and nodes in the routing graph. This algorithm can be sped up significantly during the congestion estimation mode by applying fast search techniques such as *best first search* and *A* search* [HNR68]. These techniques rely on an estimate for the distance to the destination, and are therefore not always easily applicable during the regular global routing process in which it may be difficult to estimate the cost of the unexplored portion of a route if the cost function includes components for delay or congestion. In contrast, the cost function used for routing during the congestion estimation mode is almost always the wirelength, which can be approximated at any arbitrary bin by the Manhattan distance between that bin and the destination bin.

The pseudocode in Algorithm 2 shows the general procedure for congestion estimation using a fast optimal search algorithm. It differs from the generic probabilistic congestion estimation scheme described in Algorithm 1 only in (i) the method used for the computation of routing demands, and, (ii) in avoiding any explicit modeling of the blockages. In order to determine the routing demands for a net, it applies fast search algorithms for finding the routes instead of distributing the probabilistic demands in the bounding box of the net.

Typically, in order to find a path from a source bin to a destination bin in a graph, global routers use breadth-first search either through some variant of Lee's maze routing algorithm [Lee61] or through some form of Dijkstra's shortest path algorithm. The primary limitation of these algorithms is that the paths optimizing a given cost are searched in all directions without any bias. Therefore, the search may visit a large number of bins before finding the shortest path. As an example, Fig. 2.13(b) shows the wavefront of bins being expanded in all directions while finding a path between the source and the sink pins, p_1 and p_2, depicted in Fig. 2.13(a). In this figure, the numbers in the bins represent the distance from the source. The breadth-first search begins with the source bin, visits its neighbors, and adds them to a queue;

[7] In a typical routing graph, $|E|$ is $O(|V|)$, so that the effective complexity of Dijkstra's algorithm on such graphs is $O(|V| \log |V|)$.

Algorithm 2 Congestion estimation by global routing based on fast search algorithms

1: **for all** bins B in the layout **do**
2: Compute the routing track supply for B
3: **end for**
4: Create a routing graph of bins
5: **for all** nets N in the design **do**
6: Generate a topology for N using some fast topology generation algorithm
7: **for all** segments s of the topology for N **do**
8: Use best first or A* search to find a route for s
9: **for all** bins B that lie along the route for s **do**
10: Update the routing demand in B with that arising from the route for s
11: **end for**
12: **end for**
13: **end for**
14: **for all** bins B in the layout **do**
15: Divide the routing demand for B by its available track supply
16: **end for**

the process continues until the destination bin is reached. In this example, at least twelve bins, excluding the source and the destination, must be visited before reaching the destination bin, *i.e.*, the bin containing p_2.

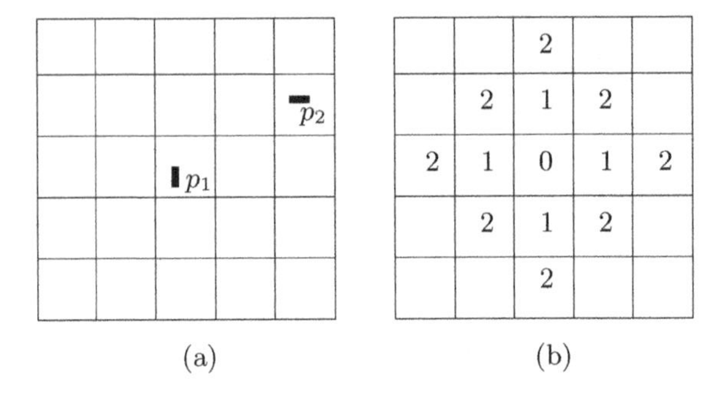

Fig. 2.13. (a) Source and sink pins in different bins. (b) Wavefront expansion using breadth-first search.

The search can be sped up by expanding the wavefront preferentially in the direction of the destination. This can be accomplished by the use of fast search techniques such as A* and best first search. A* search has been employed in the past for delay-oriented routing [PK92] (although it is slightly harder to use for this purpose with modern interconnect delay models); more recently, it has been proposed for congestion estimation purposes [WBG05].

The difference between the A* and best first search techniques is that A* considers the distance from the source as well as from the destination while

Algorithm 3 Finding a route using A* search

1: cost[s] $\leftarrow$ 0
2: Add source s to the queue Q
3: **while** Q is not empty and the destination t is not reached **do**
4: $u \leftarrow$ top[Q]
5: **if** u != t **then**
6: **for all** v adjacent to u and v is not visited yet **do**
7: cost [v] $\leftarrow$ cost [u] + 1 + estimated_distance(v, t)
8: parent[v] $\leftarrow u$
9: Insert v into Q
10: **end for**
11: Remove u from Q
12: Sort Q on cost
13: **else**
14: Trace-back the path from t to s
15: **end if**
16: **end while**
17: Return path if t is reached; otherwise, report failure

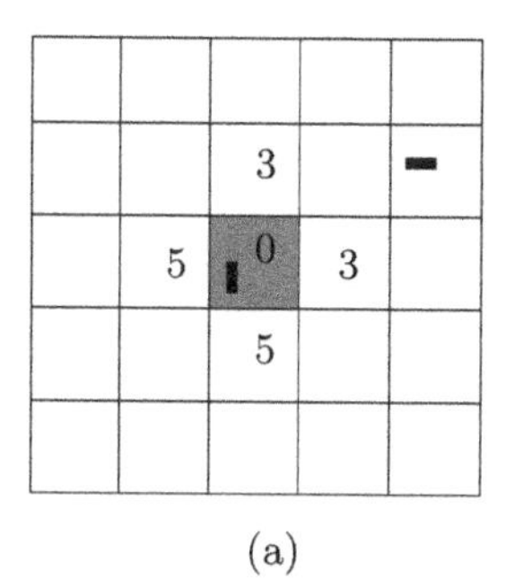

(a)

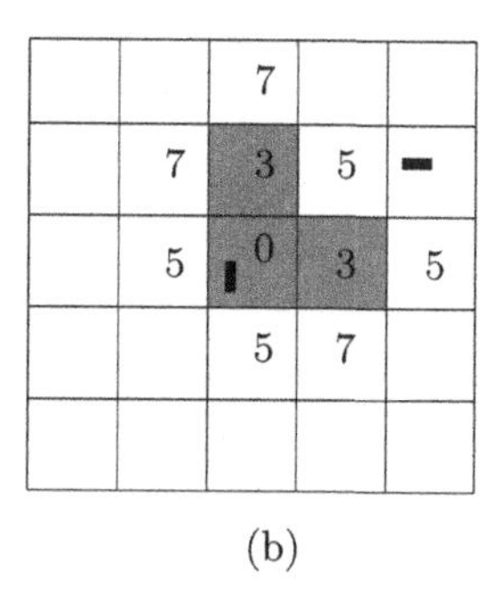

(b)

Fig. 2.14. Wavefront expansion in A* search: (a) after exploring the source bin, and, (b) after exploring the best bins from the first wavefront.

expanding the wavefront, in contrast to best first search that expands only at a bin with the best distance remaining to the destination. Note that one can only estimate the distance remaining to the destination (until the routing has completed, finding a path to the destination in the process). At any given bin, the Manhattan distance to the destination is a natural choice for the estimation of the remaining distance.

The pseudocodes for route finding with A* and best first search are shown in Algorithms 3 and 4, respectively. The resulting wavefronts due to the application of these procedures on the previous example (depicted in Fig. 2.13(a)) are shown in Figs. 2.14 and 2.15, respectively. In these figures, a number in any given bin indicates the estimated distance of the best route from the source to the destination through that bin in the case of A* search, and the estimated remaining distance from the given bin to the destination bin in the case of best first search. At each stage, bins adjacent to the shaded bins are added to the wavefront.

One can observe that both A* and the best first search find the destination bin after visiting fewer bins than plain breadth-first search. As compared to the twelve bins visited by the breadth-first search, A* search visits nine bins and the best first search visits seven bins before reaching the destination. Thus, although the asymptotic time complexity for all three search algorithms is the same (namely, $O(|E|+|V|\log|V|)$, when implemented using a Fibonacci heap), the two fast search schemes visit far fewer bins than the breadth-first search, resulting in a significant speed-up in practice.

Algorithm 4 Finding a route using best first search

1: Add source s to the queue Q
2: **while** Q is not empty and the destination t is not reached **do**
3: $u \leftarrow$ top$[Q]$
4: **if** u != t **then**
5: **for all** v adjacent to u and v is not visited yet **do**
6: cost $[v] \leftarrow$ estimated_distance(v, t)
7: parent$[v] \leftarrow u$
8: Insert v into Q
9: **end for**
10: Remove u from Q
11: Sort Q on cost
12: **else**
13: Trace-back the path from t to s
14: **end if**
15: **end while**
16: Return path if t is reached; otherwise, report failure

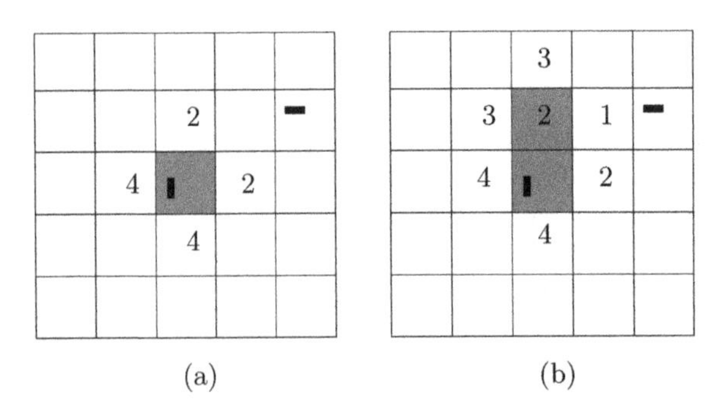

Fig. 2.15. Wavefront expansion in best first search: (a) after exploring the source bin, and, (b) after exploring a best bin from the first wavefront.

2.4 Comparison of Fast Global Routing with Probabilistic Methods

Fast global routing techniques can predict the congestion more accurately than probabilistic congestion estimation methods, but tend to be somewhat slower. A comprehensive quantitative comparison of some specific implementations of fast global routing and probabilistic congestion estimation methods has been reported in [WBG05].

Probabilistic congestion estimation is known to be pessimistic, especially when it does not include post-processing to consider detours or to model rip-up and reroute, as it is not congestion-aware. As a result, the maximum congestion, total track overflows, or the number of congested bins predicted by probabilistic estimation are overestimates for the corresponding post-routing metrics; these metrics may cause a circuit to be deemed unroutable even if it can be routed successfully. In such cases, the placement optimizations using these estimates may ignore the routable design implementation and attempt to optimize it further, which may or may not improve its routability. Fast global routing, on the other hand, tends to overestimate the congestion to a much lesser degree, since it finds routes that avoid congested bins. Consequently, it distributes the congestion evenly in a manner similar to the behavior of real routers. Although probabilistic congestion maps can be post-processed to reduce the pessimism in their prediction, this reduction comes at the cost of additional runtime.

Design blockages too are handled more naturally in fast global routing techniques than in probabilistic methods, since fast routers simply try to find a route around them just like real routers. In contrast, probabilistic methods may employ heuristics such as redistribution of the routing demands from the blocked bins, or pseudo-pin insertion. Although these heuristics may work well for simple blockages, the more complicated blockages require some form of maze routing in their vicinity, so that the probabilistic congestion estimation method starts resembling global routing on such designs. In the same vein, fast global routing based methods automatically handle detoured routes in congested regions in the process of routing the nets, whereas probabilistic methods usually do not model detours.

Typically, the probabilistic congestion estimation method based on the assumption of single and double bend routes is two to three times faster than the fast global routing technique. Probabilistic congestion prediction employing multibend routes has higher runtimes than the one based on single and double bend routes, since the routing demands for bins inside the bounding box vary significantly and require the use of binomial coefficients for their computation. Therefore, the runtime difference between the probabilistic congestion estimation method based on multibend routes and the fast global routing may not be significant. Furthermore, the use of only single and double bend routes is a better approximation of the behavior of real routers, at least on designs that are not very congested.

The asymptotic complexity to compute the routing demands for a given net using the probabilistic method is $O(b)$, whereas the same for the fast global routing technique is $O(b \log b)$, where b is the number of bins in the entire routing area (assuming the same tessellation of the routing area for the two methods). Even though the asymptotic time complexity is worse for fast global routing than for the probabilistic technique, the use of fast search algorithms such as A* or best first search often improves the runtime of fast global routing in practice. Furthermore, if post-processing techniques are employed in probabilistic estimation methods to improve the accuracy (at the cost of a runtime overhead), the difference in overall runtimes between probabilistic methods and fast global routing may no longer be significant.

Because of the relatively good accuracy that is achievable with these techniques, both probabilistic estimation and fast global routing have been used for congestion prediction in commercial physical synthesis tools.

2.5 Final Remarks

In this chapter, we reviewed several placement-level metrics for routing congestion that can be used to drive layout optimizations. These metrics vary in their accuracy, fidelity, the required computation time, and their typical application. The simple, crude but computationally fast metrics are used to guide the placement during early stages, whereas the more accurate but computationally expensive metrics are used to alleviate the congestion during the later stages of placement or to decide when to proceed to the routing stage. Although it is already quite fast, the computation of simple metrics can sometimes be sped up even further by the use of structural information such as Rent's rule, enabling the application of such metrics within the inner loops of placement-based circuit optimizations.

In contrast, probabilistic and fast global routing based schemes require much more computation runtime, since they attempt to model the behavior of real routers. However, they use several techniques to reduce the runtime as compared to real global routers, trading off accuracy for speedup. This allows them to be invoked repeatedly during the later stages of placement (although not at a granularity as fine as that possible with the simple metrics). They generate a high resolution congestion map with a good degree of accuracy, although they tend to suffer from pessimism in their predictions; the problem of pessimism is especially acute during probabilistic congestion estimation. In general, probabilistic and fast global routing based methods tend to do well in sparsely congested designs, but can be more error-prone in congested regions (where the behavior of the global router diverges from its generic behavior more significantly).

Many of these congestion estimation techniques have also been extended for use with layout-aware netlist optimizations, as will be discussed in the next chapter.

References

[CZY+99] Chen, H.-M., Zhou, H., Young, F. Y., Wong, D. F., Yang, H. H., and Sherwani, N., Integrated floorplanning and interconnect planning, *Proceedings of the International Conference on Computer-Aided Design*, pp. 354–357, 1999.

[Che94] Cheng, C.-L. E., RISA: Accurate and efficient placement routability modeling, *Proceedings of the International Conference on Computer-Aided Design*, pp. 690–695, 1994.

[Dav98] Davis, J. A., De, V., and Meindl, J. D., A stochastic wire length distribution for gigascale integration (GSI) – Part I: Derivation and validation, *IEEE Transactions on Electron Devices* 45(3), pp. 580–589, March 1998.

[Dij59] Dijkstra, E. W., A note on two problems in connection with graphs, *Numerische Mathematik* 1, pp. 269–271, 1959.

[Don79] Donath, W. E., Placement and average interconnection lengths of computer logic, *IEEE Transactions on Circuits and Systems* 26(4), pp. 272–277, April 1979.

[HNR68] Hart, P. E., Nilsson, N. J., and Raphael, B., A formal basis for the heuristic determination of minimum cost paths, *IEEE Transactions on System Science and Cybernatics* SSC-4, pp. 100–107, 1968.

[Hig69] Hightower, D. W., A solution to line routing problems on the continuous plane, *Proceedings of the Design Automation Workshop*, pp. 1–24, 1969.

[HM02] Hu, B., and Marek-Sadowska, M., Congestion minimization during placement without estimation, *Proceedings of the International Conference on Computer-Aided Design*, pp. 739–745, 2002.

[HS01] Hu, J., and Sapatnekar, S., A survey on multi-net global routing for integrated circuits, *Integration – the VLSI Journal* 31(1), pp. 1–49, Nov. 2001.

[KX03] Kahng, A. B., and Xu, X., Accurate pseudo-constructive wirelength and congestion estimation, *Proceedings of the International Workshop on System-level Interconnect Prediction*, pp. 61–68, 2003.

[LR71] Landman, B. S., and Russo, R. L., On a pin versus block relationship for partitions of logic graphs, *IEEE Transactions on Computers* C-20(12), pp. 1469–1479, Dec. 1971.

[Lee61] Lee, C. Y., An algorithm for path connection and its applications, *IRE Transactions on Electronic Computers* EC-10(3), pp. 346–365, Sep. 1961.

[LTK+02] Lou, J., Thakur, S., Krishnamoorthy, S., and Sheng, H. S., Estimating routing congestion using probabilistic analysis, *IEEE Transactions on Computer-Aided Design of Integrated Circuits and Systems* 21(1), pp. 32–41, Jan. 2002.

[PK92] Prasitjutrakul, S., and Kubitz, W. J., A performance-driven global router for custom VLSI chip design, *IEEE Transactions on Computer-Aided Design of Integrated Circuits and Systems* 11(8), pp. 1044–1051, Aug. 1992.

[SL01] Saxena, P., and Liu, C. L., Optimization of the maximum delay of global interconnects during layer assignment, *IEEE Transactions on Computer-Aided Design of Integrated Circuits and Systems* 20(4), pp. 503–515, April 2001.

[SPK03] Selvakkumaran, N., Parakh, P. N., and Karypis, G., Perimeter-degree: A priori metric for directly measuring and homogenizing interconnection complexity in multilevel placement, *Proceedings of the International Workshop on System-level Interconnect Prediction*, pp. 53–59, 2003.

[SY05] Sham, C., and Young, E. F. Y., Congestion prediction in early stages, *Proceedings of the International Workshop on System-level Interconnect Prediction*, pp. 91–98, 2005.

[Str01] Stroobandt, D., *A Priori Wire Length Estimates for Digital Design*, Norwell, MA: Kluwer Academic Publishers, 2001.

[WBG04] Westra, J., Bartels, C., and Groeneveld, P., Probabilistic congestion prediction, *Proceedings of the International Symposium on Physical Design*, pp. 204–209, 2004.

[WBG05] Westra, J., Bartels, C., and Groeneveld, P., Is probabilistic congestion estimation worthwhile?, *Proceedings of the International Workshop on System-level Interconnect Prediction*, pp. 99–106, 2005

[YKS02] Yang, X., Kastner, R., and Sarrafzadeh, M., Congestion estimation during top-down placement, *IEEE Transactions on Computer-Aided Design of Integrated Circuits and Systems* 21(1), pp. 72–80, Jan. 2002.

3

SYNTHESIS-LEVEL METRICS FOR ROUTING CONGESTION

Building congestion awareness into any given stage of a design flow requires metrics to quantify congestion estimates during that stage in order to discriminate between the congestion impact of various optimization choices. There is an inherent conflict between the accuracy and extent of the layout information available during a design stage and the level of flexibility available during that stage to modify the circuit to alleviate congestion problems. Stages that are further upstream in the design flow usually have more flexibility available to fix congestion problems, but have to deal with larger errors in congestion prediction. Congestion metrics that are appropriate to a particular design stage may not be equally applicable to other design stages, since the amount of layout information available during different stages is different.

Chapter 2 discussed several metrics used to estimate the routing congestion at the placement level, when the netlist is fairly well-established. Most of these metrics relied extensively on the locations of the cells. However, such detailed location information is not available during the technology mapping and (technology-independent) logic synthesis stages that precede placement. During these stages, the precise structure of the netlist has not yet been finalized, and the placement of the nodes in the netlist (loosely corresponding to cells) is either non-existent or approximate. On the other hand, these steps offer a great deal of freedom to alter the netlist to ameliorate congestion hot spots. Unlike placement, where the set of wires in the netlist is fixed, these steps can absorb wires within logic gates or split logic functions into smaller gates, providing additional degrees of freedom. Therefore, it is useful to also add congestion-awareness to this part of the design flow. These optimizations can be guided only by congestion metrics that can operate under greater uncertainty than placement-level congestion metrics. This chapter provides an overview of such metrics. Some of them extend the metrics at the placement level to the synthesis level, whereas others exploit graph theoretic properties of the netlist in the absence of any placement information.

3.1 Motivation

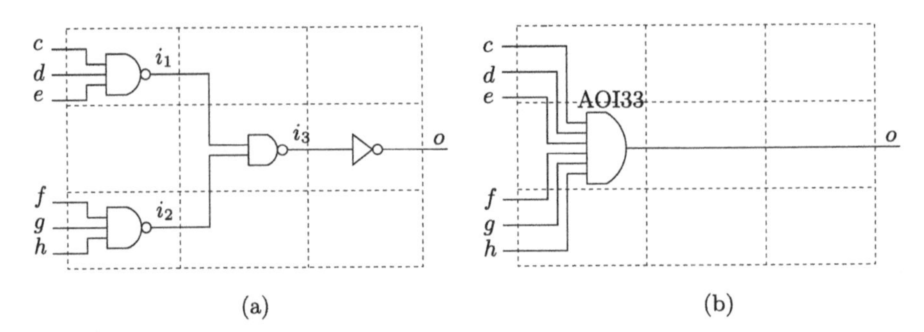

Fig. 3.1. Synthesis choices and routing demand: the suboptimal network in (a) may have smaller routing demand than its area- and delay-optimal equivalent in (b). (Reprinted from [SSS+05], ©2005 IEEE).

As an example, consider the two logically equivalent implementations of the and-or-invert (AOI) functionality shown in Figs. 3.1(a) and 3.1(b). The layout is tessellated into global routing bins, represented by dashed boxes in the figures. Assume that both the implementations receive the primary inputs c, d, and e from the top left bin, and the primary inputs f, g, and h from the bottom left bin, whereas the primary output, o, of the circuit leaves from the central bin on the right. Furthermore, let us assume that all primary inputs arrive at the same time, and that the total number of transistors that lie on the pull-up or pull-down chains on the worst delay path in an implementation is a metric for the delay of that implementation.

The implementation in Fig. 3.1(a) has nine literals in all, and the corresponding Boolean equations are shown below:

$$\begin{aligned} i_1 &= \overline{cde}, \\ i_2 &= \overline{fgh}, \\ i_3 &= \overline{i_1 i_2}, \\ o &= \overline{i_3}. \end{aligned}$$

This implementation requires eighteen transistors for its realization using static CMOS logic. For this circuit, the worst-case delay path goes through five transistors. An alternative implementation of the circuit is shown in Fig. 3.1(b), where the Boolean equation can be written as:

$$o = \overline{cde + fgh}.$$

This alternative realization requires twelve transistors, and the most critical path contains three transistors. Clearly, this alternative realization is better in terms of both gate area and delay, and a conventional technology mapper is likely to choose this realization. However, if we add congestion considerations to the picture, we find that the former choice is likely to have better routability than the latter. Assume that each internal (horizontal or vertical) bin boundary permits only one track to be used for the nets in the implementation. Then, the former implementation is routable, whereas the latter one cannot be routed irrespective of the placement of the AOI33 cell. If we measure congestion as the track demand in terms of the number of bin-boundary crossings, then the former realization has a demand of twelve, whereas the latter has a demand of at least fifteen. Note that the track demand of fifteen in the latter case corresponds to the best possible placement; if the AOI33 cell were to be placed in the second or third column, the track demand would have been even higher.

The above example highlights several observations about congestion-aware synthesis [SK01, PPS03, SSS+05, LAE+05], namely:

- Logic synthesis choices that are area- or delay-optimal may not be optimal from the point of view of congestion or routing demand, and may even lead to unroutable circuits.
- Congestion optimization requires the use of additional metrics that measure congestion, since the literals or gate-area alone cannot capture the routing demand.
- Area- or delay-optimal synthesis solutions may be unroutable even with the best possible placement, thus pointing to the futility of relying only on placement (and subsequent routing) to alleviate the congestion.

This inability of the traditional synthesis cost functions and metrics to capture the routability and congestion of an implementation of a circuit necessitates the development of new metrics for congestion that can be used in conjunction with the traditional metrics during synthesis. Any such congestion metric should have the following two capabilities:

- Given two logically equivalent netlists, the metric should be able to distinguish between them on the basis of routability.
- Given a netlist, the metric should be capable of guiding synthesis optimizations to improve its routability.

At the same time, the metric should be fast to compute. It should be capable of performing the above functions without actually going through the time-consuming steps of placement and routing. This ease of computation is essential for any such metric to be used extensively during logic synthesis transformations, since these transformations are applied repeatedly during traditional logic synthesis [SSL+92].

For the purposes of this chapter (and the remainder of this book), we distinguish between *technology-independent logic synthesis* and *technology mapping*. In a departure from the common practice of considering both steps to be ingredients of logic synthesis, we use the phrase "logic synthesis" to refer specifically to technology-independent logic synthesis, and refer to technology mapping explicitly where needed. This distinction between (technology-independent) logic synthesis and technology mapping is relevant because the congestion estimation metrics proposed for the two stages are quite different from each other. Most modern physical synthesis flows interleave the process of technology mapping with that of the placement of the evolving netlist. Although some placement-oblivious congestion metrics have been proposed for use during technology mapping, any partial placement information available at that stage can help considerably in the fast estimation of congestion (as discussed in Section 3.2). Unfortunately, this placement information is usually either not available or very inaccurate during logic synthesis. Therefore, most of the congestion metrics proposed for use during logic synthesis rely exclusively on structural features of the netlist. Congestion metrics targeted towards logic synthesis are discussed in Section 3.3.

3.2 Congestion Metrics for Technology Mapping

The goal of the technology mapping stage has traditionally been to map the logic network on to a given cell library, with the aim of obtaining a good tradeoff between area, delay, and power considerations. In typical design flows, technology mapping is carried out subsequent to the decomposition of the network into a *subject graph* consisting of primitive gates such as two-input NAND gates and inverters, and comprises of the pattern matching and covering of this subject graph using gates from the library. Technology mapping is typically implemented using some dynamic programming based procedure whose principles are rooted in [Keu87].

Conventionally, technology mapping has been guided by wire-load models, as illustrated in Fig. 3.2(a). A wire-load model is a table that lists, for different fanouts, the average netlength and capacitive load obtained after the placement of a mapped netlist. Although such approaches sufficed for older technologies that did not have significant wire resistances and wire delays, their effectiveness has been decreasing with each successive process technology node since the 250 nm node. In designs at today's technology nodes, wire-load models can be highly erroneous (because two different wires having the same fanout can have dramatically different wire delays because of their differing netlengths), and can result in numerous iterations between the mapping and placement stages without any guarantee of convergence or design closure [SN00, GOP+02].

To get around the inadequacy of wire-load models, modern technology mapping algorithms often employ placement information to guide the mapping choices [Dai01, SK01, LJC03]. This can be achieved either through the use of a *companion placement* during mapping, or through iterations between

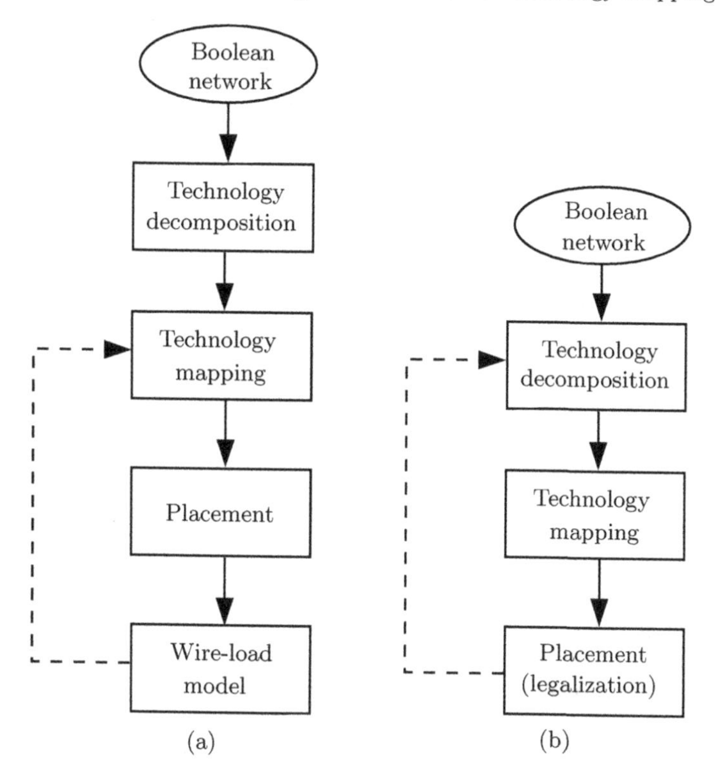

Fig. 3.2. A typical technology mapping flow (a) using wire-load models, and (b) using placement information.

mapping and placement. Although the concept of associating placement information with the primitive gates during the mapping that evolves into a companion placement for the mapped netlist has been around for more than a decade [PB91], it has gained wider acceptance only in the last few years. Alternatively, technology mapping and placement can be performed iteratively, as in [LJC03], so that the netlengths, and therefore the capacitive loads and delays of the nets, are estimated with improved accuracy in each successive iteration. A typical modern physical synthesis flow is shown in Fig. 3.2(b).

Since the placement information is already extracted for net delay calculation by modern physical synthesis flows as described above, it is natural to apply the same information to compute routing congestion also. This can be achieved by extending the congestion metrics at the placement level to compute routing congestion during the technology mapping optimizations. Indeed, many of the congestion-aware technology mapping approaches in the literature [SK01, PPS03, SSS+05, SSS06] extend placement-level congestion metrics such as netlength and probabilistic congestion maps to guide mapping choices. At the same time, a few *a priori*[1] congestion estimation schemes such as mutual contraction [HM02] have also been used to drive congestion-aware technology mapping.

3.2.1 Total Netlength

As observed in the previous chapter, the total netlength of the nets in a design correlates well with the average congestion. Therefore, minimizing the netlength is often consistent with the goal of improving the congestion. However, since it is a gross global metric, it cannot capture the spatial and locality aspects of routing congestion. In other words, the total netlength cannot predict what the congestion will be in a particular bin, or where the bins with high local congestion will be located. Furthermore, just like all other placement-based congestion estimation schemes, the netlength metric can be somewhat inaccurate if the placement assumed during the mapping is substantially different from the final placement. Another major problem with this metric is that it does not capture the discrete relationship between routability and congestion. As an example, the nets in a design may be routable when the congestion is less than some threshold, but may become unroutable or require significant detours (along with the associated delay penalties) when that threshold is crossed. Unfortunately, this behavior is not captured by the total netlength metric.

The major advantage of the netlength metric is that it is easy to compute, requiring $O(n)$ time[2] for a net with n pins. This efficiency of computation makes it an attractive choice that does not affect the asymptotic computational complexity of technology mapping algorithms. Consequently, this metric has been used in several congestion-aware technology mapping approaches [SK01, PPS03]. These approaches apply the netlength metric to reduce congestion by combining it with other traditional objectives such as area and delay. However, they achieve limited success in congestion mitigation, primarily due to the inherent limitations of this metric, as discussed above. Further details on these optimization techniques can be found in Chapter 6.

---

[1] Schemes used for individual or statistical netlength prediction that do not rely on any placement information are referred to as *a priori* schemes.

[2] For nets with more than two terminals, a minimum spanning tree (MST) estimate, which requires $O(n^2)$ time for a net with n pins, may be used. Although this is usually more accurate than the "half-rectangle perimeter of bounding box" (HRPM) estimate for the wirelength of a given net, most technology mapping approaches either use the bounding box estimate for the net (often accounting for the pin count of the net using compensation factors as discussed in Section 2.1.1), or decompose the net into two-pin nets using a star or clique model, because of the dominance of other sources of errors such as the inaccuracies inherent in the placement information at the mapping stage.

3.2.2 Mutual Contraction

The mutual contraction metric was originally proposed to make placement more congestion-aware [HM02], and has also recently been extended to congestion mitigation during technology mapping [LM05]. Mutual contraction is an *a priori* metric that uses the structure of the netlist in the topological neighborhood of a net to predict its final netlength, without relying on any placement information. At an intuitive level, it measures the tendency of the endpoints of a net to resist being pulled apart because of their connectivity to other cells. Nets whose endpoints are only weakly connected to other cells tend to have a large mutual contraction value. Since all reasonable placement engines try to place strongly connected cells together, the value of this metric for a net correlates well with the expected netlength for that net, with large mutual contraction values corresponding to short netlengths.

Given a circuit graph $G = (V, E)$ for a netlist, where the vertices in V correspond to the cells of the netlist and the edges in E are used to model the nets, the mutual contraction metric can be defined on each of the edges in E. For the purpose of this metric, multipin nets are modeled using cliques. Thus, a net connecting n cells is modeled using $n(n-1)/2$ edges corresponding to all possible pairs among the n pins of the net. For simplicity, let us consider the case of traditional placement that minimizes wirelength. In this case, all nets are weighted equally in the cost function for the placement (with the weight of each net being, say, one). The weight of a net is distributed equally among all the edges that are used to model that net. Thus, the contribution of an edge $e(u, v)$ from a clique denoting a connection between n vertices that includes u and v to the weight $w(u, v)$ is given by $2/(n(n-1))$. Note that an edge $e(u, v)$ can simultaneously belong to several different cliques (corresponding to different multipin nets that share two or more pins). In such a case, the total weight of an edge is the sum of the contributions from each of the cliques that contain that edge.

The *relative weight* $w_r(u, v)$ of the edge $e(u, v)$ is now defined as the ratio of the weight $w(u, v)$ to the sum of the weights of all the edges incident on u. Observe that although $w(u, v) = w(v, u)$ for any edge $e(u, v) \in E$, the relative weights $w_r(u, v)$ and $w_r(v, u)$ may not be the same.

Definition 3.1. *The mutual contraction of an edge $e(u, v)$ is defined as the product of relative weights $w_r(u, v)$ and $w_r(v, u)$.*

In other words, the mutual contraction $mc(e(u, v))$ for edge $e(u, v)$ is given by:

$$mc(e(u,v)) = \frac{(w(u,v))^2}{\sum_{z:e(z,u)\in E} w(z,u) \sum_{z:e(z,v)\in E} w(z,v)}. \tag{3.1}$$

The example shown in Fig. 3.3 illustrates the computation of the mutual contraction metric. Assuming that each edge represents a single fanout, the relative weight $w_r(u, v)$ for the edge $e(u, v)$ in Fig. 3.3(a) is given by:

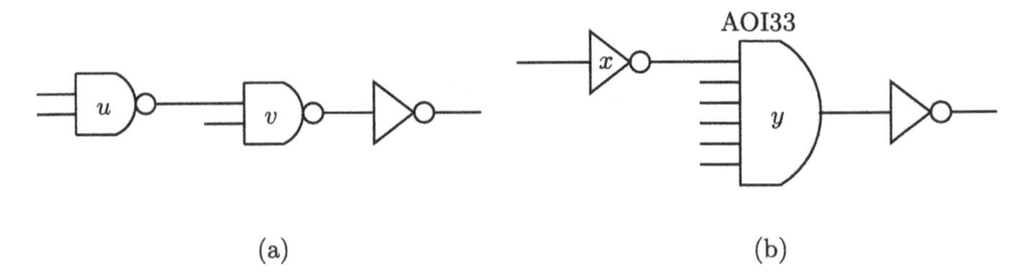

Fig. 3.3. The mutual contraction for the edges $e(u,v)$ in (a) and $e(x,y)$ in (b) is 1/9 and 1/14, respectively.

$$w_r(u,v) = \frac{w(u,v)}{\sum_{z:e(z,u)\in E} w(z,u)} = \frac{1}{3}.$$

Similarly, $w_r(v,u)$ is also 1/3. Therefore, the mutual contraction for the edge $e(u,v)$ is given by:

$$mc(e(u,v)) = w_r(u,v) \times w_r(v,u) = \frac{1}{9}.$$

Similarly, in Fig. 3.3(b), the relative weights $w_r(x,y)$ and $w_r(y,x)$ can be easily computed as 1/2 and 1/7 respectively, resulting in a mutual contraction of 1/14 for the edge $e(x,y)$.

The mutual contraction metric can be used to estimate the netlength associated with an edge; the greater the value of the metric, the smaller is the expected separation between the two cells corresponding to the endpoints of the edge in the final placement. In the above example, since $e(u,v)$ has greater mutual contraction than $e(x,y)$, it is also likely to have the shorter netlength. The intuitive justification for this correlation can be explained in the context of the example. The number of edges that compete against a short netlength for $e(u,v)$ is four, these nets being the net driven by v and the three fanin nets of the cells u and v other than the net (u,v). However, the corresponding number of competing edges for the edge $e(x,y)$ is seven. Therefore, the cells u and v are likely to be placed closer to each other than the cells x and y.

As with all structural metrics that ignore placement information, the mutual contraction metric is not very effective at predicting the netlength for any given net with high accuracy. However, the work in [HM02] empirically demonstrates a high negative correlation[3] (typically in the $(-0.9, -0.6)$ range) between the mutual contraction and the average netlength for nets at the placement level. In other words, the higher the mutual contraction for an

[3] A statistical correlation value that is close to +1.0 (−1.0) means that the corresponding variables are strongly positively (negatively) correlated, whereas a correlation value close to zero implies that the corresponding variables are independent of each other.

edge, the shorter is its expected netlength. Since the total netlength is a measure of the average congestion, employing mapping choices with higher mutual contraction is likely to lead to a netlist that has a shorter total netlength. This has motivated the congestion-aware technology mapping work in [LM05]. However, although this work has shown some promising results in congestion mitigation, the use of this metric suffers from the same problems as the use of the total netlength to measure congestion, in addition to the inaccuracies inherent in the use of placement-oblivious structural netlength prediction metrics. Furthermore, the mutual contraction metric is not very effective at predicting the netlength for multipin nets. Therefore, technology mapping based on mutual contraction proposed in [LM05] requires the application of additional metrics such as the *net range* (discussed in Section 3.3.3).

Computationally, mutual contraction is more expensive than netlength computations. The computation of the mutual contraction of an edge $e(u, v)$ requires $O(deg(u) + deg(v))$ time, where $deg(u)$ and $deg(v)$ correspond to the degrees of the nodes u and v. In contrast, the netlength computation for a two-pin net can be performed in constant time.

Although mutual contraction is the only *a priori* structural metric that has been used for congestion-aware technology mapping to date, it is worth pointing out that several other such metrics have been proposed for total and individual netlength prediction. These metrics are briefly discussed later in Section 3.3.5. Among these metrics, mutual contraction has been shown in [LM04] to exhibit better correlation with average netlengths than *connectivity* and *edge separability*. However, the recently proposed *intrinsic shortest path length* metric [KR05] yields even better correlation than the mutual contraction metric, and is likely to be a good candidate for congestion-aware technology mapping driven by structural metrics.

3.2.3 Predictive Congestion Maps

An alternative approach to estimating routing congestion is the use of post-placement probabilistic congestion maps to guide technology mapping. However, as illustrated in Fig. 3.4(a), this approach poses a "chicken-and-egg" problem in conventional design flows because the congestion map is available only after the placement of the mapped netlist, whereas the placement step requires a solution from the technology mapping step. Predictive congestion maps have been proposed in [SSS+05] as a way to overcome this problem.

The construction of these predictive congestion maps is illustrated in Fig. 3.4(b). It involves the use of the subject graph, which is a netlist containing only primitive gates such as two-input NANDs and inverters on which the technology mapping is performed. The subject graph, which is also known as a *premapped netlist*, is placed within a specified block area. Next, this placement is used to construct the probabilistic congestion map by applying the probabilistic estimation methods involving either unlimited bends or L- and Z-shapes for wires [LTK+02, WBG04] (discussed in Section 2.2 in Chapter 2)

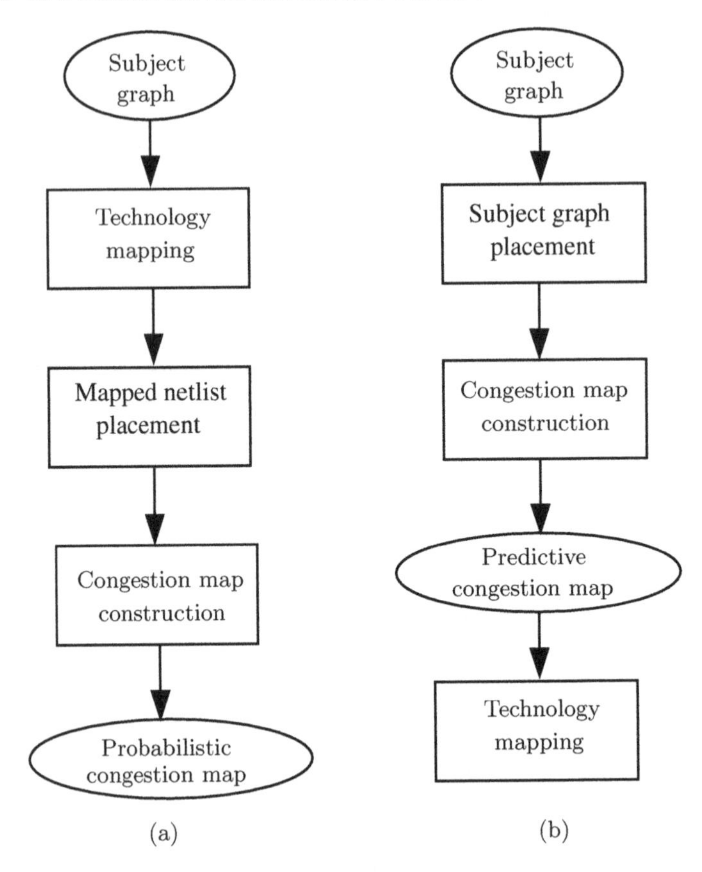

Fig. 3.4. Construction of (a) conventional and (b) predictive congestion maps.

to the edges in the placed subject graph. The resulting congestion map is referred to as a *predictive congestion map.* The work in [SSS+05] showed that there is a good correlation[4] between the congestion map predicted prior to technology mapping and actual congestion map obtained after placement.

The intuitive justification for this correlation is that the premapped and mapped netlists share the same global connectivity. This is because technology mapping merely subsumes a subset of wires in the subject graph as internal connections within the standard cells, thus retaining the similarity in the connectivity between the premapped and the post-mapped netlists. Since the connectivity and other constraints such as block area and I/O locations are the same, most reasonable placement algorithms (including partitioning-based approaches such as [CKM00] as well as analytical methods such as [EJ98], to

[4] A statistical correlation of greater than 0.6 between the predicted and actual congestion maps has been reported in [SSS+05] across different circuit families, logic synthesis scripts, technology mapping algorithms, and placement techniques.

be discussed briefly in Section 5.1 in Chapter 5) generate placements with similar congestion distributions for both the netlists.

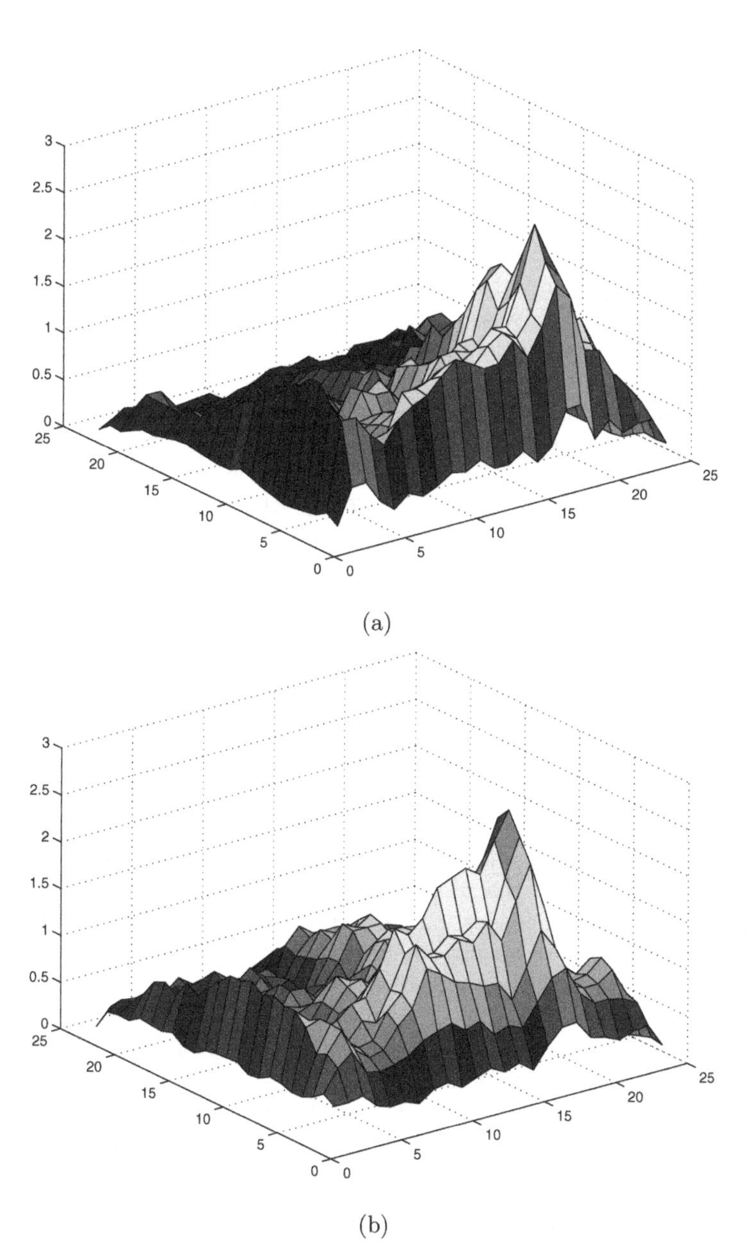

Fig. 3.5. Horizontal congestion of a typical benchmark circuit: (a) congestion map for the mapped netlist (optimized for gate area), and, (b) congestion map for the premapped netlist. (Reprinted from [SSS+05], ©2005 IEEE).

As an example, Figs. 3.5(a) and (b) depict the horizontal congestion maps for the mapped and premapped netlists, respectively, for a benchmark circuit consisting of an arithmetic logic unit (ALU) and some associated control logic. (The vertical congestion maps, omitted for brevity, exhibit a similar correlation). In these figures, the horizontal (XY) plane denotes the block area and the vertical (Z) axis shows the congestion. This circuit has been technology-mapped using an industrial library used in high performance microprocessor design at the 90 nm process technology node, and the placement of premapped as well as mapped netlists has been carried out using an industrial force-directed placer. The similarity in congestion maps is readily observable, even though the placer is allowed to use a pin-bloating technique to alleviate local congestion. In particular, one can see that the location of the peak congestion correlates quite well across the two congestion maps, even though this location is offset from the center of the block (that tends to show the peak congestion in most designs).

In this scheme, multipin nets are first decomposed into a set of equivalent two-pin nets using a clique or star model prior to estimating their contribution to the congestion map. The worst-case time complexity for the computation of congestion contribution for a two-pin net is $O(b)$, where b is the number of bins in the entire layout area (since the bounding box of the net may span the entire layout area in the worst case). Thus, predictive congestion maps are computationally more expensive than the netlength or mutual contraction metrics.

As with the netlength metric, another overhead associated with the computation of a predictive congestion map is the cost of placing the subject graph. Subject graphs typically contain many more nodes than the mapped netlists for the corresponding circuits, so that the cost of placing them may also be much higher than that of placing the final netlist. However, the legalization of this placement may be omitted without any adverse effects. Indeed, a case may be made that continuing the placement of the subject graph to a granularity finer than that of the tessellation of the layout area provides no additional accuracy to the congestion map, which is discretized at the granularity of the bins. This coarse placement is usually much faster than a full-fledged placement of the subject graph. Any errors due to cell overlaps in this placement are anyway likely to be dominated by the inaccuracies introduced due to the displacement of the cells relative to the placement of the subject graph when the mapped netlist is placed.

Another consequence of the fact that the subject graph typically contains many more nodes than the final netlist is that it requires a larger area for a legal placement than the final netlist. This necessitates a scaling of the dimensions of the primitive gates corresponding to the nodes in the subject graph, in order to fit the subject graph into the layout area for the final netlist. A simple way to carry out this scaling is to decrease the cell area for the nodes in the subject graph by a factor equal to the ratio of the expected area of an implementation based on the subject graph to the area available for the final layout.

Predictive congestion maps suffer from a few sources of errors. If the placement techniques applied after the technology mapping produce a significantly different placement, then the predicted congestion mitigation achieved during technology mapping based on the predictive congestion map may not materialize. Moreover, the predictive congestion map is independent of the mapping algorithm, since it is generated purely on the basis of the subject graph. This insensitivity of the predictive congestion map to the actual mapping solution can introduce additional errors in the congestion prediction. The constructive congestion maps discussed next in Section 3.2.4 avoid this source of error.

In general, in spite of these errors, predictive congestion maps, as well as the constructive congestion maps to be introduced in the next section, capture a much richer level of detail in the spatial distribution of estimated congestion in the final placement than *a priori* structural metrics or the total netlength. This spatial information can be exploited by technology mapping algorithms to choose between conventional or congestion-aware modes, so that the gate area or delay overheads associated with congestion-optimal choices occur only where absolutely necessary [SSS+05]. In contrast, structural and netlength metrics do not offer these flexibilities since they are oblivious to spatial and locality information about congestion. Therefore, technology mapping algorithms that use congestion maps are typically more effective at congestion optimization than those that rely on structural or netlength metrics.

3.2.4 Constructive Congestion Maps

Although predictive congestion maps provide a coarse level of accuracy in congestion prediction, they have some inherent limitations. Specifically, predictive congestion maps are unable to discriminate between the congestion maps corresponding to different technology mapping solutions obtained from the same subject graph, since each such solution would be associated with an identical map. This problem is remedied by the use of *constructive congestion maps* [SSS06] that retain the level of spatial detail provided by predictive congestion maps while increasing its accuracy.

Constructive congestion maps are created dynamically and propagated during the technology mapping process. The essential idea behind the generation of these maps is best illustrated by an example. Figure 3.6(a) shows a small subject graph in which a match M_1 is being considered for a node N_1. The technology mapping step is performed by extending the conventional matching and covering phases with an additional step that determines the congestion map corresponding to each candidate solution. Thus, whenever a node is mapped, the congestion maps associated with all of its predecessors are known and can be used to generate the congestion map associated with that node. For our example, the congestion maps due to solutions at N_2 and

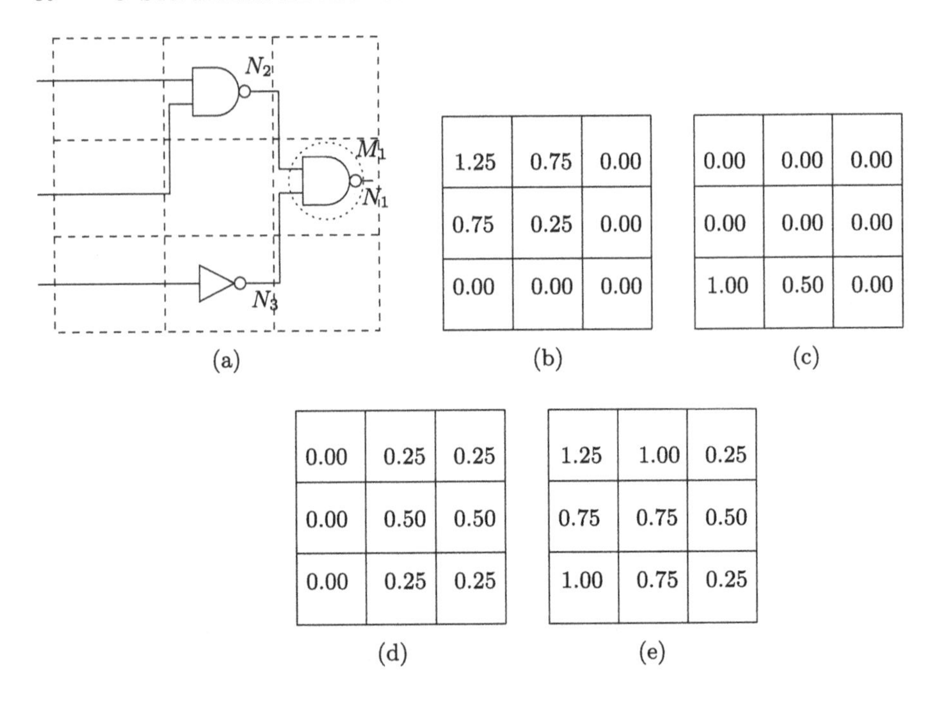

Fig. 3.6. Constructive congestion map generation: (a) Subject graph during matching process in technology mapping with match M_1 at node N_1. (b) Horizontal track demand due to solutions at N_2. (c) Horizontal track demand due to solutions at N_3. (d) Horizontal track demand due to fanin nets to the match M_1. (e) Overall horizontal track demand due to the mapping solution because of M_1. (Reprinted from [SSS06], ©2006 IEEE).

N_3 are as shown in Fig. 3.6(b) and Fig. 3.6(c), respectively. Only the horizontal routing demand is shown in our example for the sake of simplicity. These congestion maps are constructed probabilistically [LTK+02, WBG04], as discussed in Section 2.2 in the previous chapter.

For the circled match M_1 at node N_1, where a two-input NAND gate is chosen, Fig. 3.6(d) shows the congestion map associated with this match only, accounting for the nets due to two fanins of the match. This computation is straightforward; as an example, the bins in the top-right corner in Fig. 3.6(d) have a track demand of 0.25, since only one out of the two possible routes passes through it and this requires only half a track. Note that the bins in the first column in the figure show zero routing demand values, since the bounding boxes of the nets that feed the match M_1 do not contain those bins. Finally, Fig. 3.6(e) shows the congestion map representing the mapping solution due to the match M_1 at N_1. It is obtained by the bin-wise addition of the congestion maps in Figs. 3.6(b), 3.6(c), and 3.6(d).

The congestion map for the mapping solution due to a match is propagated across a multifanout point by distributing the congestion equally among all the fanouts. The matching process continues in topological order, so that the congestion maps due to all the wires in the mapping solutions are available at the primary outputs when the process finishes. As with the predictive congestion map, the constructive map also requires the subject graph to be placed prior to the technology mapping. Whenever a new node is generated during the matching process, it can be placed at the center of gravity of the nodes that it subsumes (since their locations are already known at that stage). Note that different mapping solutions represent different set of wires. Unlike predictive congestion maps that are static regardless of the match that is chosen, the constructive map captures the congestion impact of different matches dynamically.

The worst-case time complexity for the computation of the constructive congestion map for a two-pin net is the same as that of the predictive one, namely, $O(b)$, where b is the total number of bins for the entire layout area. Constructive maps, however, require more memory than predictive maps, since different (partial) congestion maps due to selected matches at different primary outputs may be required to create these maps for the final mapping solution. Specifically, the constructive congestion maps require $O(bn_M^{po})$ memory, where n_M^{po} is the sum of the number of different matches at all the primary outputs. There are several heuristics suggested in [SSS06] that can help in substantially reducing this memory overhead.

3.2.5 Comparison of Congestion Metrics for Technology Mapping

The main features of the routing congestion metrics for technology mapping discussed so far are summarized in Table 3.1. The metrics are listed in Column 1 and their major properties are shown in Column 2. For each metric, the time complexity for the computation of the value of the metric for a two-pin net (u, v) is listed in Column 3; in this column, $deg(x)$ represents the degree of the node x, whereas b denotes the number of bins in the layout.

Except for mutual contraction, all of the remaining metrics discussed above depend on placement information. While they tend to be better at predicting congestion than purely structural metrics like mutual contraction, they can suffer from inaccuracies if the initial placement does not reflect the final placement well. This can sometimes be a problem in physical synthesis flows that iterate between technology mapping and placement. In contrast, there are other flows that allow the placement of the subject graph or the netlist during the early stages of mapping to evolve into the final placement, using techniques of incremental placement and legalization to map netlist changes into the layout. In such flows, the congestion gains obtained using the placement-based metrics are likely to be retained.

Metric	Properties	Time complexity
Netlength	• Placement dependent • Correlated with average congestion • Does not capture spatial aspects	Constant
Mutual contraction	• Placement independent • Correlated with mean netlengths • Does not capture spatial aspects	$deg(u) + deg(v)$
Predictive congestion map	• Placement dependent • Captures spatial aspects • Considers subject graph nets	$O(b)$
Constructive congestion map	• Placement dependent • Captures spatial aspects • Considers actual nets	$O(b)$

Table 3.1. A comparison of routing congestion metrics for technology mapping.

Mutual contraction or other structural metrics are best used when the placement of the mapped netlist is likely to be very different from any placement of preliminary versions of the netlist. In such cases, there is little correlation between the predictive or constructive congestion maps and the actual congestion map obtained after the final placement. However, structural metrics suffer from their own sources of inaccuracies, as discussed earlier.

The netlength metric is easy to compute, but employing it to alleviate congestion requires changing the constants in the cost function $K_1 \times Area + K_2 \times Delay + K_3 \times Netlength$ during the mapping procedure, depending on the severity of congestion; however, this severity is not known until the mapping procedure is complete [PPS03]. Furthermore, this metric cannot discriminate between congested and uncongested regions in a layout. Therefore, the (usually small) regions of peak congestion determine the severity of the gate area and delay penalty for the entire netlist. In contrast, predictive and constructive congestion maps overcome this limitation and allow the selection of congestion-optimal matches in congested regions and area-optimal or delay-optimal matches in uncongested regions. Constructive maps are the most accurate of all these metrics, since they capture the congestion due to only the relevant set of wires that actually exist in the final netlist, and exclude the effects of the wires that are absorbed within the mapped gates.

Typically, the computational complexity as well as the memory requirement of any congestion estimator increases with the desired accuracy and effectiveness. For a two-pin net, the netlength metric requires constant memory and computation time independent of the routing of the net. However, the runtime of an overall congestion-aware flow using this approach may be more than that of a conventional flow that ignores congestion constraints, since the placement of the subject graph requires additional runtime. Employing mutual contraction for technology mapping does not require subject graph placement, but the time complexity of the mapping algorithm worsens, since the computation of this metric for a two-pin net is more expensive than that of the netlength. Even then, this timing complexity remains independent of the routing of the net. In practice, the runtime penalty of congestion-aware technology mapping using mutual contraction is seen to be negligible [LM05].

The overhead for the subject graph placement generation applies to all placement dependent metrics, including predictive and constructive congestion maps. The time complexity for computing the contribution of a two-pin net to either of these congestion maps depends on the span of the net (*i.e.*, its bounding box), since large, spread-out nets impact a larger number of placement bins. Constructive maps provide the best accuracy, but require considerably more memory than the other approaches (although this memory limitation may be largely overcome by the use of efficient heuristics).

A study of how these metrics are employed within congestion-aware technology mapping algorithms is presented in Sections 6.2.3 and 6.2.4 in Chapter 6.

3.3 Routing Congestion Metrics for Logic Synthesis

Technology-independent logic synthesis has traditionally aimed at minimizing the number of literals or the number of logic levels in a multilevel Boolean network obtained from the register transfer level (RTL) description of a design, since these metrics have historically correlated well with the area and delay, respectively, of the final implementation of the network. Although this correlation is not always good, especially with the increasing dominance of interconnect delays in modern process technologies, these metrics sufficed in the era when the gate delay dominated the overall delay of the circuits (namely, in technologies with feature sizes greater than 250 nm). Furthermore, poor interconnect scaling is resulting in a large increase in the number of buffers required for the alleviation of poor delays and slews in resistive wires [SMC+04]. In such a scenario, these traditional metrics for logic synthesis are no longer as effective as in the past. The number of literals merely captures the number of transistors required for the static CMOS implementation of the specified logical functionality as discussed in Section 3.1, ignoring the transistors required for the buffers necessary to achieve the expected performance. In addition, although the transistor area does correlate well with the leakage power dissipation, it is not necessarily a good indicator of the total design area in wire-limited designs that may require considerable "white space" (as will be discussed in Chapter 5). Similarly, the number of logic levels at the logic synthesis stage has become inaccurate as a delay metric, since it does not capture the delay of buffered interconnects or the impact of sizing the gates. The development of alternative reliable metrics to drive logic synthesis, however, requires further research. Therefore, the number of literals and the number of logic levels are still used as proxies for area and delay, respectively, in many of today's logic synthesis tools.

There have been several attempts to develop interconnect-aware metrics for logic synthesis. Most of these have been placement-oblivious, relying instead on the structural properties of various nets in the netlist to predict the behavior of the downstream mapping and layout tools on those nets. Placement-aware techniques such as the use of iterations between synthesis and placement or of a companion placement, that have proven so effective at the technology mapping stage in modern physical synthesis flows, are usually not as successful when applied to logic synthesis. Several works [GNB+98, SK01, CB04] have attempted to use placement information during logic synthesis, but have had limited success for the following reasons:

- The predictive accuracy and fidelity of the placement of a netlist keeps decreasing as one moves farther upstream in the synthesis flow. While the congestion map errors are still tolerable at the technology mapping stage, the errors in the congestion values of placement bins at the logic synthesis stage are often of the same order of magnitude as or even larger than the congestion values themselves, because of which the congestion maps can be very misleading at this stage. Indeed, the impact of the synthesis transformations at this stage is often so large that it is difficult to maintain any consistency in the placement information.
- The nodes in the Boolean network available during logic synthesis usually show much greater variance in the area required by them in the final implementation than nodes in the subject graph or the partially mapped netlist. This is because RTL is usually written from a logical perspective, with little attention being paid to the layout aspects of the Boolean network. In contrast, nodes in the subject graph are quite uniform in area.
- The nature of the subsequent technology mapping is such that only a subset of nets from the network optimized at the logic synthesis stage appear in the mapped netlist, since many of the nets are subsumed as internal connections inside the cells.

Consequently, there does not seem to be much benefit in capturing the routing congestion predicted at the logic synthesis stage in a two-dimensional congestion map. In contrast, several simpler metrics that can help discriminate between two different implementations of a netlist in terms of their likely netlength and congestion have been studied for use during logic synthesis. These metrics rely on structural properties of graphs, and include the literal count, adhesion [KSD03], fanout range [VP95], net range [LM05] and neighborhood population [HM96b]. Other metrics such as edge separability [CL00], closeness [SK93], connectivity [HB97], and intrinsic shortest path length [KR05], which also rely on graph structure and are typically used to drive the placement towards minimizing the total netlength or for *a priori* netlength prediction, can also be extended to guide the synthesis transformations.

3.3.1 Literal Count

As discussed previously, the traditional area metric used at the logic synthesis stage is the number of literals. It has been observed that when a Boolean network is mapped on to a trivial library containing only two-input NAND gates and then placed using a partitioning-based algorithm, there is a strong linear relationship between the number of literals and the peak congestion [KSD03]. Intuitively, the number of literals correlates with the number of nets in the circuit; in two equivalent Boolean network representations of a given circuit, the network with the smaller number of literals is likely to have fewer nets and therefore, potentially better routability.

However, as pointed out earlier, the number of literals does not represent all the nets that will be in the circuit, since the technology mapping stage decides which connections actually become wires, and which remain internal to the gates. In fact, only a subset of nets from the technology-independent representation may appear in the mapped netlist. This is because the network is first decomposed into primitive gates, followed by matching and covering with the library cells during the technology mapping. The decomposition stage introduces new nets and then the matching and covering stages subsume many of the newly created and original nets. This results in a set of nets in which only a fraction may be survivors from the technology-independent representation.

The number of literals can be computed in time that is linear in the size of a network. Traditional synthesis transformations require only the computation of the *literal gain*, which too can be measured in time that is linear in the number of affected nets. Thus, the literal count is a fast, albeit crude, metric for estimating routing congestion.

3.3.2 Adhesion

The connectivity of the graph representing a Boolean network encodes considerable information about its layout-friendliness. For instance, it is easy to see that networks that form non-planar graphs[5] are likely to result in greater routing congestion. Similarly, a more "entangled" graph is likely to have greater routing congestion than a less entangled one. An attempt to capture this notion of entanglement in Boolean networks involved a comprehensive study of several metrics such as literal count, cell count, average fanout, levelization, and adhesion [KSD03]. Among these, adhesion was found to be the most

[5] A graph is non-planar if and only if it contains a subgraph that is a subdivision of one of the two Kuratowski subgraphs K_5 (the clique on five vertices) and $K_{3,3}$ (the complete bipartite graph on six vertices partitioned into two sets of three vertices each) [Kur30] [Har94].

promising candidate for the *a priori* prediction of peak routing congestion. The remaining candidate metrics, except for the literal and cell counts, were found to exhibit poor correlation with peak routing congestion.

Definition 3.2. *The adhesion of a network is defined as the sum of the min-cuts between all pairs of nodes in the network.*

The adhesion of a network displays a good correlation with maximum congestion after mapping the network using a simple library and placing it. For a number of designs obtained from the MCNC and ISCAS benchmark suites, a plot of the adhesion versus the peak congestion has been shown to have a least square linear fit with an R^2 value[6] of 0.643 [KSD03]. The adhesion metric can be used to build congestion awareness into synthesis transformations. For example, given two equivalent Boolean networks with different adhesion values, the one with the lower adhesion would be likely to have smaller peak congestion. Adhesion can also be used to guide logic synthesis transformations such as extraction, as described in Chapter 6.

We have seen that the adhesion metric does not capture the routing congestion completely by itself. Another possible metric is a weighted sum of the adhesion, the number of literals, and the number of cells. The value of R^2 for a linear fit between the peak congestion and this new metric has been found to be significantly better than that with the adhesion metric alone (although the experiments assumed that a simple library containing only two-input NAND gates is used for the mapping) [KSD03]. Indeed, the usefulness of this metric for logic synthesis followed by technology mapping using realistic libraries is currently unknown.

The computation of adhesion is expensive, since an exact algorithm requires $O(nF(n, m))$, where $F(n, m)$ is the time required for solving the maximum flow problem in a network with n vertices and m edges [ACZ98]. The Edmonds-Karp implementation of the Ford-Fulkerson algorithm to solve the maximum flow problem requires $O(|V||E|^2)$ time, where $|V|$ ($|E|$) is the cardinality of the set of vertices (edges) in the network [EK72]. More efficient algorithms such as [GT88] still require $O(|V||E|\log(|V|^2/|E|))$ time. Thus, the overall time complexity is at least cubic in the number of nodes. With approximation algorithms, it is possible to compute adhesion in time that is linear in the size of the network with some error. Even with these approximations, the linear time complexity of adhesion makes it an expensive overhead within synthesis transformations.

[6] The R^2 metric, also known as the *coefficient of determination*, measures the fraction of the variance in an observed data set that can be explained by a regression. More formally, given an observed set of n data points, $\{(x_1, y_1), (x_2, y_2), \cdots, (x_n, y_n)\}$, and the regression line $\hat{y_i} = a + bx_i$ obtained by, say, least square fitting, the R^2 metric is computed as $\frac{\sum_{i=1}^{n}(\hat{y_i} - \overline{y})^2}{\sum_{i=1}^{n}(y_i - \overline{y})^2}$, where $\overline{y}$ is the mean over the observed values of y_i. The closer the value of this metric is to 1.0, the better is the fit [Mee99].

3.3.3 Fanout and Net Range

The fanout range metric has been proposed as a candidate to guide the extraction process during logic synthesis to improve the routability [VP95], and has also been extended further to create a new congestion metric called the net range [LM04]. Assuming that the depth of each node in a circuit graph is computed using a topological traversal, these metrics are defined as follows:

Definition 3.3. *The fanout range of a node is the range of circuit depths (i.e., the number of topological levels) spanned by its fanout terminals.*

Definition 3.4. *The net range of a node is the range of circuit depths spanned by all of its terminals.*

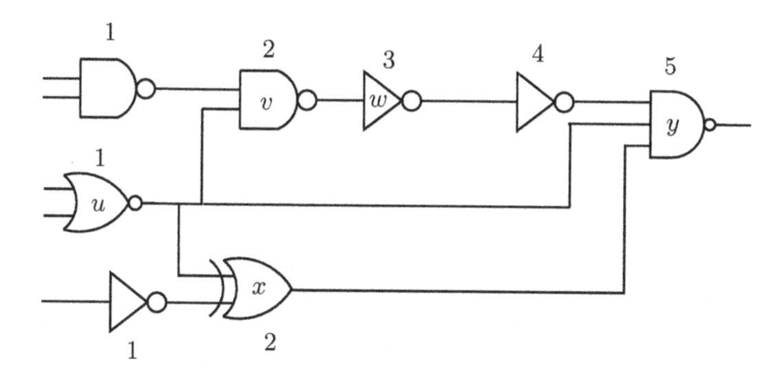

Fig. 3.7. An example of fanout and net range computation.

Figure 3.7 shows an example of fanout and net range computations for a circuit. In this figure, the depth associated with each cell is shown next to it. For instance, the depth of cell u is one, whereas that of y is five. The fanout range of the net driven at the output of gate u is three, since the minimum depth of any of its fanouts is two (corresponding to cells v and x), whereas the maximum is five (corresponding to y); the difference between these is the fanout range. The net range for the same net, however, considers the driver u as well, resulting in a value of four. It is obvious that for single fanout net, the fanout range is always zero, whereas the net range has a non-zero value that depends on the depth of the driver and receiver. For instance, the fanout range is zero for each of the two-pin nets (v, w) and (x, y). However, the net range values for these nets are one and three, respectively. Thus, the net range is more discriminatory than the fanout range because, unlike the latter metric, it can also identify nets whose drivers are topologically distant from their receivers, potentially leading to long routes.

The fanout and net range metrics try to capture, in a graph theoretic sense, how far the fanouts or all cells connected by a net are likely to be placed, in the absence of any actual placement information. Intuitively, the greater the fanout or net range, the larger is the expected netlength, since the corresponding cells are likely to be placed far apart. Of course, this is not strictly true, since, in real circuits, even two-pin nets with small net range values can correspond to long wires, especially if they do not lie on any critical path. Reducing the fanout or net range also implies that the connections are localized merely in a topological sense; the actual netlength or routing congestion may not be reduced in all cases.

The time complexity for the computation of either of these metrics for a net is linear in the degree of the net. Therefore, it can be easily incorporated into various cost functions during synthesis. The fanout range metric has been used to guide fast extraction procedures during logic synthesis to improve the wirelength in [VP95], whereas the net range metric has been used for fanout optimization in [LM05].

3.3.4 Neighborhood Population

Another way to measure the local routing congestion caused by a cell is to measure the number of topologically connected cells that are likely to be placed in its vicinity, since the nets connected to a cell compete for routing resources with nets of all the other cells in its neighborhood. The neighborhood population metric attempts to capture this expected congestion in a graph theoretic sense, using the notion of the *distance* between two cells, defined as the number of cells on the shortest path between them. This distance is computed on the undirected graph that underlies the (directed) circuit graph, so that pairs of cells that do not have any path passing through both of them can contribute to the distance metric for one another. Note that this notion of distance is different from that of the depth ranges for different cells that was discussed in Section 3.3.3; computation of the depth ranges relied only on edges in the directed circuit graph.

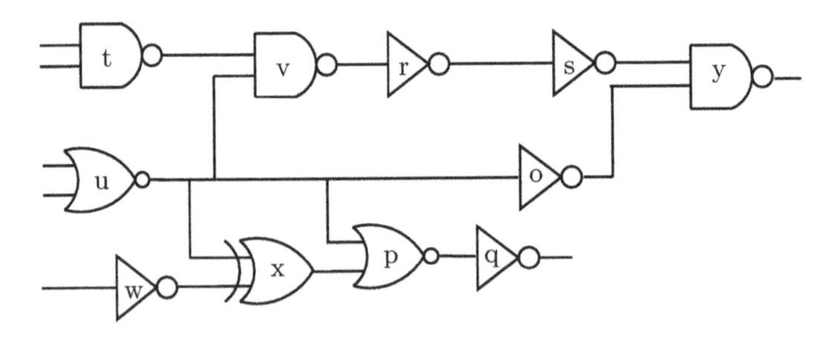

Fig. 3.8. An illustration of the neighborhood population metric.

As an illustration of the computation of the distance metric, consider the example in Fig. 3.8. In this example, let us ignore the inputs and outputs of the circuit for the sake of simplicity. Then, there are two cells at a distance of one from cell s, namely, cells v (across r) and o (across y). Similarly, there are six cells at a distance of one from cell r, namely, y (across s) and t, u, x, p and o (all across v).

The neighborhood population metric was first introduced to estimate the wirelength and layout area for logic netlists implemented in a two metal layer process technology [PP89a, PP89b]. It was further parameterized based on distance and used to predict the yield on metal layers as a function of total estimated netlength and critical area for a given netlist without actually performing any layout [HM96b, HM96a, HKM+97]. While the neighborhood population metric had originally been defined for nets, it was subsequently extended to cells and applied to routability enhancement.

Definition 3.5. *The neighborhood population at a distance i for a given cell c, $Ngh(c)_i$, is the number of cells at a distance i from the cell.*

Thus, for the example in Fig. 3.8, $Ngh(s)_1 = |\{v, o\}| = 2$ and $Ngh(r)_1 = |\{t, u, x, p, o, y\}| = 6$.

The neighborhood population definition can be further extended to compute the total and average neighborhood populations over any subcircuit by, respectively, summing and taking the average of the corresponding values over all cells in the subcircuit.

The notion of the neighborhood population has been used to modify the cost functions in logic synthesis transformations such as *substitution* in an attempt to improve the routability of the circuit [KK03]. This work also presents some evidence that the application of other transformations such as *fast extraction* and *speed-up* [SSL+92] also results in netlists with differing values for the neighborhood population metric, which is expected to correspond to different levels of congestion in the final layout.

The computation of the neighborhood population metric is more expensive than that of the fanout and net range metrics. Specifically, if the metric is used as a congestion discriminator between two logically equivalent netlists being explored during synthesis, it requires a traversal of the entire network to compute the total and average neighborhood populations. Even during local synthesis optimizations, this metric takes longer to compute than the fanout or net range estimations, since its value depends on all the nodes within distance i, and not just on the fanin and fanout nodes.

3.3.5 Other Structural Metrics for Netlength Prediction

In this section, we will briefly review some other metrics that have been proposed either for the *a priori* estimation of individual netlengths or the average netlength in a design, or for the purpose of guiding placement algorithms to generate more routable designs. While these metrics have not been used during logic synthesis to date, they are candidates for such an application just like the other structural metrics discussed earlier. These metrics include edge-separability [CL00], connectivity [HB97], closeness [SK93] and intrinsic shortest path length [KR05]. Each of these metrics attempts to capture the tendency of tightly connected groups of cells or clusters to be placed close to each other.

Definition 3.6. *The edge-separability for an edge $e(u, v)$ in graph $G(V, E)$ is defined as the minimum cut size among the cuts separating u and v.*

For example, the edge-separability for the edge $e(v, r)$ in the example depicted in Fig. 3.8 is two, since edges $e(v, r)$ and $e(r, s)$ form a minimum cut-set separating v and r. Similarly, the edge separability for the edge $e(t, v)$ is one.

Definition 3.7. *The intrinsic shortest path length (ISPL) of a net $e(u, v)$ in the graph $G(V, E)$ is defined as the shortest path length between u and v in the graph $G'(V, E - \{e(u, v)\})$.*

In the above example, $ISPL(e(r, s))$ is four, corresponding to the path $r - v - o - y - s$.

Definition 3.8. *The closeness between two cells or clusters of cells is defined as the ratio of number of nets connecting them to the minimum of the degree of either of the two cells or clusters.*

Continuing further with our example of Fig. 3.8, one can verify that the closeness between cells v and r is $1/2$, since the number of edges connecting v and r is one, while $\min(deg(v), deg(r)) = \min(6, 2) = 2$. Note that the net (u, v, x, p, o) is modeled using a clique, resulting in edges between all pairs of cells connected by the net.

Definition 3.9. *The connectivity between two cells or clusters of cells u and v is given by:*

$$connectivity_{uv} = \frac{bw_{uv}}{size_u \times size_v \times (fo_u - bw_{uv}) \times (fo_v - bw_{uv})}$$

where bw_{uv} is the sum of weights of all edges between u and v, $size$ is the measure of the area of a cell or cluster, and fo represents the degree of a cell or cluster.

This metric attempts to capture the relative strengths of the tendency of two cells or clusters to be pulled together because of the net(s) connecting them and their tendency to be pulled away from each other because of the nets connecting them to their other neighbors. In our ongoing example, if the size of each cell and the weight of each edge is one, the connectivity of the edge $e(r, s)$ is one (because both r and s each have only one additional neighbor).

There has been some work [HM02] that demonstrates that mutual contraction is a better metric than connectivity or edge separability in terms of its ability to discriminate between short and long nets. More recently, [KR05] has shown that the ISPL metric outperforms both mutual contraction and edge separability in terms of the correlation between the predicted and actual netlengths (although it is comparable to the mutual congestion metric in its ability to predict which of a pair of nets will be longer). Therefore, it is a promising candidate for use in congestion-aware logic synthesis and technology mapping.

3.3.6 Comparison of Congestion Metrics for Logic Synthesis

In summary, the metrics that have been used to estimate routing congestion at the logic synthesis stage include the number of literals, adhesion, fanout and net range, and neighborhood population. The literal count metric is computationally the least expensive. However, it does not capture congestion well, although it is better at estimating the cell area. The adhesion metric is the most expensive of all the proposed metrics, and is best used to capture the routing congestion in conjunction with additional metrics such as the number of literals and the number of cells. The correlation of these metrics with the peak congestion has been demonstrated after technology mapping using a trivial library, but is unknown for realistic libraries that include many complex gates. The neighborhood population metric is less expensive than adhesion and has shown some promise when used with synthesis transformations. The fanout and net range metrics are computationally less expensive than even the neighborhood population. Unlike the net range metric, fanout range cannot distinguish between short and long two-pin nets. Net range has shown encouraging results when applied to fanout optimization after technology mapping.

However, the correlation of all these metrics has been studied only against average netlengths, and therefore, average congestion. It is unclear whether these metrics will be able to predict local congestion hot spots. It is, thus, apparent that the metrics for routing congestion at the logic synthesis stage require further research to be truly effective for synthesis optimizations.

3.4 Final Remarks

In this chapter, we have reviewed several routing congestion metrics for technology mapping and logic synthesis. Technology mapping often employs placement dependent metrics such as netlength and congestion maps, although graph theoretic measures such as mutual contraction can also be applied. Among these, constructive congestion maps are the most accurate and effective, especially when the assumed placement is preserved, but are computationally expensive. In contrast, the netlength is computationally the least expensive, but is also inaccurate and therefore, less effective. Furthermore, as with mutual contraction and the structural metrics proposed for congestion estimation during the logic synthesis stage, the netlength metric suffers from the inability to identify locally congested regions.

All the congestion metrics that have been applied at the logic synthesis stage are graph theoretic measures, since the logic synthesis transformations cause such large perturbations to the network structure that maintaining consistent placement information is a challenge. Among the metrics such as literals, adhesion, fanout and net range, and neighborhood population, it is not yet clear which has the best correlation with the post-mapping and post-placement routing congestion for realistic designs and libraries. Indeed, there is still much scope for further research in the quest for a good congestion metric applicable at the logic synthesis stage.

References

[ACZ98] Arikati, S., Chaudhuri, S., and Zaroliagis, C., All-pairs min-cut in sparse networks, *Journal of the ACM* 29(1), pp. 82–110, 1998.

[CKM00] Caldwell, A. E., Kahng, A. B., and Markov, I. L., Can recursive bisection alone produce routable placements?, *Proceedings of the Design Automation Conference*, pp. 477–482, 2000.

[CB04] Chatterjee, S., and Brayton, R., A new incremental placement algorithm and its application to congestion-aware divisor extraction, *Proceedings of the International Conference on Computer-Aided Design*, pp. 541–548, 2004.

[CL00] Cong J., and Lim, S., Edge separability based circuit clustering with application to circuit partitioning, *Proceedings of the Asia and South Pacific Design Automation Conference*, pp. 429–434, 2000.

[Dai01] Dai, W., Hierarchial physical design methodology for multi-million gate chips, *Proceedings of the International Symposium on Physical Design*, pp. 179–181, 2001.

[EK72] Edmonds, J., and Karp, R. M., Theoretical improvements in the algorithmic efficiency for network flow problems, *Journal of the ACM* 19, pp. 248–264, 1972.

[EJ98] Eisenmann, H., and Johannes, F. M., Generic global placement and floorplanning, *Proceedings of the Design Automation Conference*, pp. 269–274, 1998.

[GT88] Goldberg, A., and Tarjan, R. E., A new approach to the maximum-flow problem, *Journal of the ACM* 35(4), pp. 921–940, 1988.

[GOP+02] Gopalakrishnan, P., Odabasioglu, A., Pileggi, L. T., and Raje, S., An analysis of the wire-load model uncertainty problem, *IEEE Transactions on Computer-Aided Design of Integrated Circuits and Systems* 21(1), pp. 23–31, January 2002.

[GNB+98] Gosti, W., Narayan, A., Brayton, R. K., and Sangiovanni-Vincentelli, A. L., Wireplanning in logic synthesis, *Proceedings of the International Conference on Computer-Aided Design*, pp. 26–33, 1998.

[HB97] Hauck, S., and Borriello, G., An evaluation of bipartitioning techniques, *IEEE Transactions on Computer-Aided Design of Integrated Circuits and Systems* 16(8), pp. 849–866, August 1997.

[Har94] Harary, F., Planarity, *Graph Theory*, Reading, MA: Addison-Wesley, pp. 102–125, 1994.

[HKM+97] Heineken, H., Khare, J., Maly, W., Nag, P., Ouyang, C., and Pleskacz, W., CAD at the design-manufacturing interface, *Proceedings of the Design Automation Conference*, p. 321, 1997.

[HM96a] Heineken, H., and Maly, W., Standard cell interconnect length prediction from structural circuit attributes, *Proceedings of the Custom Integrated Circuits Conference*, pp. 167–170, 1996.

[HM96b] Heineken, H., and Maly, W., Interconnect yield model for manufacturability prediction in synthesis of standard cell based designs, *Proceedings of the International Conference on Computer-Aided Design*, pp. 368–373, 1996.

[HM02] Hu, B., and Marek-Sadowska, M., Congestion minimization during placement without estimation, *Proceedings of the International Conference on Computer-Aided Design*, pp. 739–745, 2002.

[Keu87] Keutzer, K., DAGON: Technology binding and local optimization by DAG matching, *Proceedings of the Design Automation Conference*, pp. 341–347, 1987.

[KK03] Kravets, V., and Kudva, P., Understanding metrics in logic synthesis for routability enhancement, *Proceedings of the International Workshop on System-level Interconnect Prediction*, pp. 3–5, 2003.

[KSD03] Kudva, P., Sullivan, A., and Dougherty, W., Measurements for structural logic synthesis optimizations, *IEEE Transactions on Computer-Aided Design of Integrated Circuits and Systems* 22(6), pp. 665–674, June 2003.

[Kur30] Kuratowski, C., Sur le problème des courbes gauches en Topologie, *Fundamenta Mathematica* 16, pp. 271–283, 1930.

[LAE+05] Lembach, R., Arce-Nazario, R. A., Eisenmenger, D., and Wood, C., A diagnostic method for detecting and assessing the impact of physical design optimizations on routing, *Proceedings of the International Symposium on Physical Design*, pp. 2–6, 2005.

[LJC03] Lin, J., Jagannathan, A., and Cong, J., Placement-driven technology mapping for LUT-based FPGAs, *Proceedings of the International Symposium on Field Programmable Gate Arrays*, pp. 121–126, 2003.

[LM04] Liu, Q., and Marek-Sadowska, M., Pre-layout wire length and congestion estimation, *Proceedings of the Design Automation Conference*, pp. 582–587, 2004.

[LM05] Liu, Q., and Marek-Sadowska, M., Wire length prediction-based technology mapping and fanout optimization, *Proceedings of the International Symposium on Physical Design*, pp. 145–151, 2005.

[LTK+02] Lou, J., Thakur, S., Krishnamoorthy, S., and Sheng, H. S., Estimating routing congestion using probabilistic analysis, *IEEE Transactions on Computer-Aided Design of Integrated Circuits and Systems* 21(1), pp. 32–41, January 2002.

[Mee99] Meershaert, M., *Mathematical Modeling*, 2nd. ed., San Diego, CA: Academic Press, 1999.

[PB91] Pedram, M., and Bhat, N., Layout driven technology mapping, *Proceedings of the Design Automation Conference*, pp. 99–105, 1991.

[PP89a] Pedram, M., and Preas, B., Accurate prediction of physical design characteristics for random logic, *Proceedings of the International Conference on Computer Design*, pp. 100–108, 1989.

[PP89b] Pedram, M., and Preas, B., Interconnection length estimation for optimized standard cell layouts, *Proceedings of the International Conference on Computer-Aided Design*, pp. 390–393, 1989.

[PPS03] Pandini, D., Pileggi, L. T., and Strojwas, A. J., Global and local congestion optimization in technology mapping, *IEEE Transactions on Computer-Aided Design of Integrated Circuits and Systems* 22(4), pp. 498–505, April 2003.

[KR05] Kahng, A. B., and Reda, S., Intrinsic shortest path length: A new, accurate a priori wirelength estimator, *Proceedings of the International Conference on Computer-Aided Design*, pp. 173–180, 2005.

[SMC+04] Saxena, P., Menezes, N., Cocchini, P., and Kirkpatrick, D. A., Repeater scaling and its impact on CAD, *IEEE Transactions on Computer-Aided Design of Integrated Circuits and Systems* 23(4), pp. 451–463, April 2004.

[SN00] Scheffer, L., and Nequist, E., Why interconnect prediction doesn't work, *Proceedings of the International Workshop on System-level Interconnect Prediction*, pp. 139–144, 2000.

[SSL+92] Sentovich, E. M., Singh, K., Lavagno, L., Moon, C., Murgai, R., Saldanha, A., Savoj, H., Stephan, P., Brayton, R., and Sangiovanni-Vincentelli, A., SIS: A system for sequential circuit synthesis, *Memorandum No.* UCB/ERL M92/41, University of California, Berkeley, CA, May 1992.

[SSS+05] Shelar, R., Sapatnekar, S., Saxena, P., and Wang, X., A predictive distributed congestion metric with application to technology mapping, *IEEE Transactions on Computer-Aided Design of Integrated Circuits and Systems* 24(5), pp. 696–710, May 2005.

[SSS06] Shelar, R., Saxena, P., and Sapatnekar, S., Technology mapping algorithm targeting routing congestion under delay constraints, *IEEE Transactions on Computer-Aided Design of Integrated Circuits and Systems* 25(4), pp. 625–636, April 2006.

[SK93] Shin, H., and Kim, C., A simple yet effective technique for partitioning, *IEEE Transactions on Very Large Scale Integration Systems* 1(3), pp. 380–386, September 1993.

[SK01] Stok, L., and Kutzschebauch, T., Congestion aware layout driven logic synthesis, *Proceedings of the International Conference on Computer-Aided Design*, pp. 216–223, 2001.

[VP95] Vaishnav, H., and Pedram, M., Minimizing the routing cost during logic extraction, *Proceedings of the Design Automation Conference*, pp. 70–75, 1995.

[WBG04] Westra, J., Bartels, C., and Groeneveld, P., Probabilistic congestion prediction, *Proceedings of the International Symposium on Physical Design*, pp. 204–209, 2004.

Part III

THE OPTIMIZATION OF CONGESTION

4

CONGESTION OPTIMIZATION DURING INTERCONNECT SYNTHESIS AND ROUTING

Traditionally, the fundamental goal of routing has been route completion, which translates directly to congestion management. Over the years, there has been considerable work on the routing problem with the goal of improving route completion on difficult test cases. As discussed in Section 1.1, the most popular routing paradigm consists of a two stage routing process. The first stage, namely, *global routing*, involves route planning at the granularity of coarsely defined regions called *global routing cells* or *bins* that ignores the actual pin hookups; the goal of this stage is to avoid or minimize routing overflows in the bins. This is followed by *track assignment* and *detailed routing*, during which pin hookups within and in the neighborhood of each bin are completed and the routing is legalized.

However, the challenges faced by routers have become more diverse in recent process generations. Since interconnect delays scale much worse than device delays, good design of signal interconnects can have a significant impact on the delay of critical paths in modern designs. Interconnect design includes not only the generation of tree topologies for multipin nets and the management of detours in the routing, but also layer assignment, wire sizing and wire spacing, as well as buffer insertion and shielding for some of the nets. The increasing number of layers in modern process technologies, the heterogeneity of these layers (in terms of the parasitics of minimum-width wire segments routed on them), and the increasing resistance of vias are making layer assignment an integral part of performance-driven routing. Similarly, the choice of appropriate wire sizes, tapers and spacings for long wires can improve their delay significantly. The increasing resistance of wires is leading to a rapid increase in the fraction of signal nets that require buffering in order to meet delay and signal slew constraints. The via stacks and routing detours required by the signal nets to access the buffers inserted within them can cause significant congestion deterioration. Furthermore, the embedding of buffers into the layout cuts down the rip-up and reroute flexibility for the buffered nets significantly. In high-end designs, many signal nets require shielding in order to control injected noise or avoid signal slowdown due to cross-coupling with neighboring nets. These shields also consume valuable routing resources. At the same time, the fraction of routing resources required for adequate power delivery is also increasing. Moreover, routers may also be required to minimize the number of irredundant vias as much as possible, since vias are typically difficult to manufacture and can lead to functional or parametric yield loss. Thus, the performance-driven routing required in modern designs involves not only the traditional responsibility of route completion, but also the complex allocation of the different routing resources in a way that the required performance constraints can be met. Furthermore, each of these allocation decisions has an impact on the routing congestion.

4.1 Congestion Management during Global Routing

As mentioned earlier, the primary goal of global routing is to plan global routes in a way that avoids (or bounds) any overflows of routing demand in the global routing cells. Although this does not fully capture the local routing complexity caused due to pin accessibility issues, good global route planning goes a long way in ensuring that the design can be routed successfully, by presenting the subsequent detailed routing phase with local switchboxes that are not too difficult, and allowing that phase to focus exclusively on local route completion issues.

In order to carry out global routing, the routing region is overlaid with a coarse grid that divides it into the routing bins (as was discussed in Section 1.1). These bins may be uniform or non-uniform in size, and the grid that creates these bins may be complete, or may omit some edges over routing blockages. The routing bins imply a routing graph, with each bin corresponding to a routing graph node that is located at the center of that bin. The nodes corresponding to two adjacent bins are connected by a routing edge whose capacity equals the number of routing tracks that cross the common boundary of the two bins. The global routing process maps the nets in the design to the edges of the routing graph, with the goal of minimizing the overflows along these edges.

The biggest challenge faced by global routers is that of net ordering, as illustrated in Fig. 4.1. This figure depicts three nets $n1$, $n2$ and $n3$, routed in a region where the horizontal capacity of each routing bin is two tracks. The circled numbers in the figure indicate the order in which the nets are routed. If net $n3$ is routed last (as shown in the upper figure), it will require a detour if the first two nets are routed as shown (since all the horizontal tracks in its row would already have been used up). In contrast, if $n3$ is routed earlier, all the nets can obtain minimum length routings (as shown in the lower figure), thus reducing the total wirelength of the design.

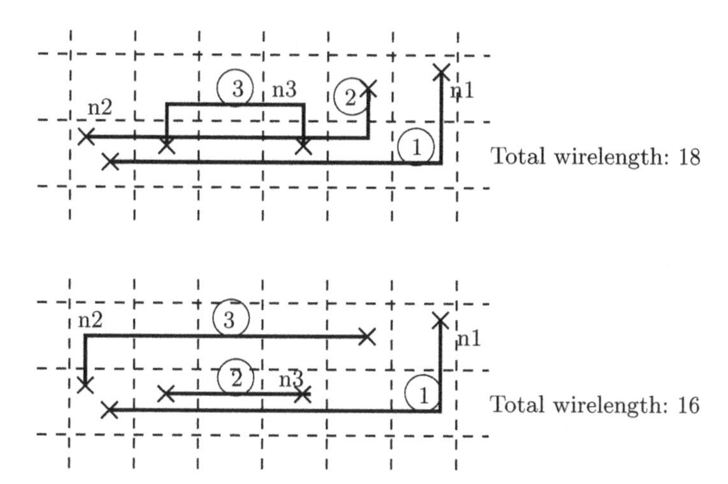

Fig. 4.1. The impact of net ordering on global routing. (The circled numbers denote the order in which the nets are routed).

Although techniques for finding a good route for a given net in the presence of a given set of blockages are well understood, the design of a scheme that can do so for all the nets in a design is still an art because of the reduced flexibility available for nets that are routed late. Thus, the design of a good global router involves getting around this net ordering problem in some way so that the nets routed late do not become the bottlenecks for route completion or performance due to excessive usage of the preferred routing resources by the nets that are routed early. Techniques that attempt to minimize or avoid the impact of net ordering on the layout quality include rip-up and reroute based schemes (discussed in Section 4.1.2), hierarchical methods (discussed in Section 4.1.3), multicommodity flow based schemes (discussed in Section 4.1.4), simulated annealing and other move-based heuristic schemes (discussed in Section 4.1.5), iterative deletion (discussed in Section 4.1.6), network flow based schemes [HS00], and dynamic area quotas [SL01]. Today's industrial global routers rely largely on heavily tuned sequential routing and rip-up and reroute algorithms, often augmented with hierarchical routing and other techniques.

In general, given the high complexity of the search space explored during global routing, the various approaches to global routing represent different tradeoffs between simplicity and computation. Thus, for example, although rip-up and reroute heuristics are usually quite simple, they can be tuned extensively to specific design styles to achieve high solution quality. In contrast, although multicommodity flow based approaches are more sophisticated and offer theoretical guarantees on solution quality, their implementations are usually more difficult to scale to very large designs or tune for particular design styles. Furthermore, these sophisticated formulations are often harder to adapt to real-world constraints such as crosstalk and non-default routing rules. This tradeoff between quality and computation also holds true within many of the commonly used global routing paradigms; permitting a rip-up and reroute or simulated annealing based global router to run longer will usually result in improved solution quality.

A good description of the various techniques commonly used for global routing can be found in [HS01] or in any of the several existing textbooks on physical design; in this section, we present merely a brief overview of these techniques.

4.1.1 Sequential Global Routing

Sequential routing is perhaps the oldest and simplest of the techniques used for global routing. It refers to the strategy of choosing some ordering for the nets, and then routing them sequentially in that order. Thus, at the time any given net is being routed, the blockages and congestion created due to nets routed earlier are known, allowing the use of any shortest path algorithm to determine the routing for the current net. Even if the current net is a multipin net, its decomposition into two-pin subnets can benefit from the available congestion map that includes the effect of all previously routed nets.

On the other hand, the biggest weakness of this technique is that the quality of the layout is very sensitive to the ordering of the nets. Although good routings are easily found for the nets routed early, the nets that are processed late can end up being unroutable or routed with very large detours or poor layer assignments. Finding a good ordering of the nets for the purpose of sequential routing is a very hard problem.

This strategy originated in an era when wire delays were insignificant, so that the detours in the routing of a net had no impact other than increasing metal track usage, thus leaving fewer routing resources for nets that were yet to be routed. In contrast, some of the nets in today's designs that have significant delays and lie on timing-critical paths cannot afford poor routings; these nets are natural candidates for early routing in any sequential scheme. However, even a previously non-critical net that is being routed late may end up with a detour or poor layer assignment, causing it to become timing-critical.

Global routers that spread routing congestion rather than greedily find the shortest possible path usually produce better quality layouts [NHL+82]. This can be achieved by the use of a dynamic weight for each edge in the underlying routing graph to model the current or predicted routing congestion along that edge. Examples of such a weight include $u(e)/s(e)$ and $(u(e)/s(e))^2$, where $u(e)$ is the current number of used tracks along routing edge e that has a total supply of $s(e)$ tracks[1]. Another weight function that is even more effective at spreading congestion (although at the cost of introducing greater routing detours) is given by $\frac{u(e)+1}{s(e)-u(e)}$ for $u(e) < s(e)$ and ∞ for $u(e) \geq s(e)$.

[1] The notions of demand and supply along the edges in a routing graph were discussed in Section 1.1.

The weight function can also have a coarser granularity, as in [CWS94] that partitions the routing space into regions using the Hanan grid[2] induced by the nodes of the current net and the corners of each (rectilinear) blockage within the routing space. Each region is assigned a weight that reflects the routing complexity and congestion within that region.

Given any weight function, the routing of a two-pin net is straightforward; the router merely finds the shortest weighted path using Dijkstra's algorithm for maze search [Lee61] or a faster but often suboptimal line probe search [Hig69]. However, routing a multipin net while spreading congestion is more complicated, since it involves the construction of a Steiner tree. Many global routing algorithms (such as [CWS94], as well as many commercial tools) use approximation algorithms that minimize the weighted wirelength of the Steiner tree for the net.

Another interesting formulation for congestion-aware Steiner tree construction uses the concept of a Steiner min-max tree (SMMT). A SMMT minimizes the maximum weight (*i.e.*, minimizing the congestion along the most congested edge) among all the edges in the tree, and can be constructed using an algorithm presented in [CS90]. However, such a tree provides no guarantees on the wirelength of the net. Therefore, [CS90] imposes explicit wirelength bounds on every SMMT. The algorithm iterates over the entire netlist in increasing order of the size of the bounding box of each net, until all the nets have been routed. The routing of nets within each iteration is controlled by a wirelength limit ratio ρ specifying the extent of wirelength degradation permissible within that iteration. More specifically, if the wirelength of the SMMT of a net N_i is greater than ρ times the semiperimeter of the bounding box of N_i, the SMMT is discarded in the current iteration. Thus, at the end of the iteration, some nets may be left unrouted. The value of ρ is initialized to a small value (usually between one and two), and relaxed in each successive iteration, so that all the nets are eventually routed. Thus, this scheme ensures that, even as the congestion is being spread, the nets that can be routed with little wirelength degradation actually do get such routings (although the total wirelength of the design may be rather large, and some critical nets may have poor routings).

4.1.2 Rip-up and Reroute

Rip-up and reroute schemes began as a feedback mechanism to ease the net ordering problem of sequential global routing. Today, these schemes are the primary workhorse of most commercial global routers. The standard approach is to route each net in a congestion-oblivious fashion, identify the congestion hot spots, and then locally reroute the segments of the nets contributing to

---

[2] The Hanan grid induced by a set of points lying in a region is the non-uniform grid generated by adding a horizontal and a vertical line extending to the boundary of the region at each of the points.

these hot spots through less congested regions. This causes the routing of a net to degrade only if the degradation is essential for congestion relief. However, rip-up and reroute can also be used to post-process any global routing to improve the overall congestion (or even individual nets that currently have poor routings or large delays).

One of the early strategies for rip-up and reroute was proposed in [TT83]. This work focuses on identifying a set of congested global routing bin boundaries, as well as a set of nets crossing these boundaries that would be ripped up and rerouted in order to relieve congestion across the selected bin boundaries. One of their observations is that if some set of congested bin boundaries forms a closed loop, then no rerouting of a net that crosses that loop exactly once can help reduce the total routing overflow along that loop. This is because at least one of the terminals of such a net lies inside the loop and at least one of its terminals lies outside the loop, as illustrated for nets $n2$ and $n5$ in Fig. 4.2. Therefore, any route for the net will have to cross the loop at least once. Only those nets that cross such loops two or more times are suitable candidates for rerouting, as is the case with nets $n1$ and $n3$ in our example.

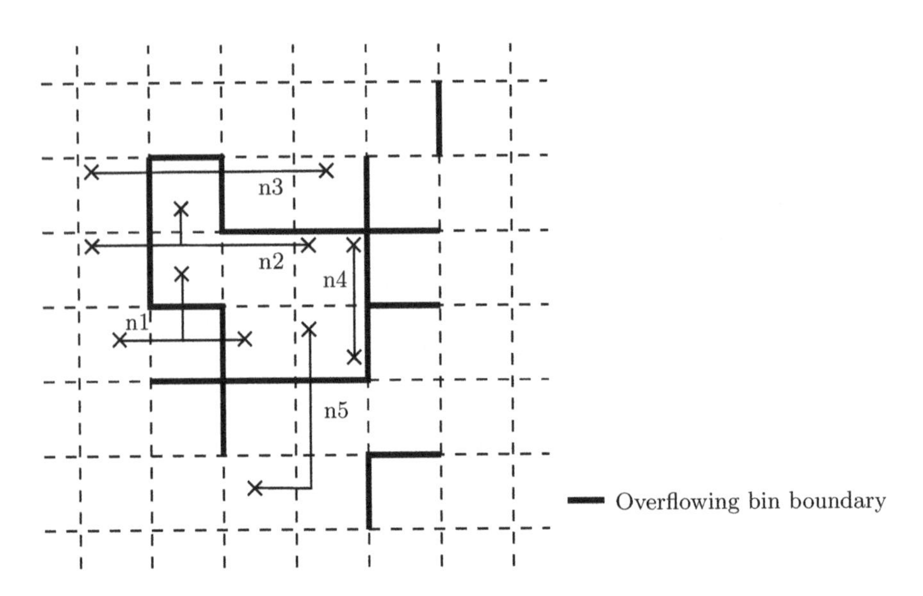

Fig. 4.2. Congestion along overflowing bin boundaries that form a loop.

Once such loops have been fixed, the algorithm of [TT83] next selects the k most congested bin boundaries. It then constructs a bipartite graph consisting of vertices corresponding to these boundaries b_i in one partition and vertices corresponding to the nets n_j crossing these boundaries in the other partition. For each such net n_j, the vertex corresponding to n_j is connected to each of the vertices corresponding to the selected bin boundaries that n_j

crosses. Next, the algorithm finds a minimum cardinality cover of the vertices corresponding to the nets that has the following property. If the routing overflow of the boundary b_i is denoted by o_i, then, for each selected boundary b_i, the number of times that the vertex corresponding to b_i is included in the cover is o_i. This yields the smallest possible set of candidates for rip-up and reroute that has the potential of resolving the congestion problems of the selected bin boundaries. This process is repeated iteratively until all overflows are eliminated from the bin boundaries, or there is no further congestion improvement.

Another similar strategy for rip-up and reroute is presented in [SK87]. Although this formulation is applicable only to two-pin nets and represents the problem as a multicommodity flow, the algorithm to select the next net for rip-up is largely independent of the network flow formulation. All nets are first routed in a congestion-oblivious fashion using a shortest path algorithm. Let ϕ be the resulting maximum overflow among all the routing edges. Then, the cost of all edges that have an overflow of ϕ or $\phi - 1$ is set to infinity. Next, each of the nets that was earlier routed along some maximum congestion edge is rerouted by a shortest path algorithm on the updated routing graph, keeping all nets unchanged except for the one being currently rerouted. The rerouting that involves the smallest cost increase is accepted, and the entire process is repeated (until no net can be rerouted with finite cost or all routing overflow is eliminated).

In contrast to the schemes described above that rip up nets passing through the most congested routing edges, the rip-up and rerouting scheme presented in [Nai87] proposes to rip-up and reroute *every* net in the same constant order iteratively, routing each net based on the (dynamically updated) congestion map available at that time. The intuition behind this approach is that the initial routing of nets selected for early routing is based on poor congestion estimates, in contrast to the routing of nets routed later. Therefore, the early selected nets should also be the ones corrected first through rip-up and reroute.

Other schemes to select routes for rip-up are ordered by the increase in the wirelength of the rerouted nets, rather than by the overflow in the routing edges. Thus, for instance, a scheme described in [LS91] orders the candidates for rip-up (*i.e.*, the nets or Steiner tree edges passing through at least one congested routing edge) by whether they can be rerouted through uncongested edges using a single bend ("L") route, followed by those that can be routed using a two bend minimum length ("Z") route, and then those that require a two bend detour (*i.e.*, a "U" route) sorted by the length of the detour. Any remaining nets that require rerouting are handled using a standard maze router.

The rip-up schemes described so far reroute entire nets (or two-pin subnets). However, this is not necessary; it is possible to design a rerouting scheme in which an alternative routing is found only for that portion of a net that currently lies along a congested routing edge. A good example of such a scheme is a network flow based rerouting described in [ML90]. For any node v in the node that acts as a super sink of the network flow. Thus, in our example, we connect the pseudo-nodes t_1 and t_2 to each of the vertices along T_1 and T_2, respectively, as well as to the super sink z. The node v serves as the source of the network flow. Each undirected edge in the routing graph is mapped to a pair of (forward and backward) directed edges with cost equal to their edge length and capacity equal to the number of routing tracks available along the undirected edge. Then, a minimum cost flow of s units is found through the network, where s is the number of selected nets. Such a flow minimizes the wirelength of the rerouting while completing all the partially routed trees without creating routing overflows. Since there are only s in-edges incident to the super sink, the flow can never be larger than s units. On the other hand, since each partial net corresponds to a single unit capacity edge incident on the super sink, a flow of less than s units implies that the routing of some partial net cannot be completed without creating an overflow.

This entire process is repeated for each node in the routing graph. Although this scheme is sensitive to the order in which nodes in the routing graph are selected, this ordering problem is usually easier than the net ordering problem. At any selected node of the routing graph, the partial trees belonging to different nets are completed concurrently, thus avoiding the net ordering problem locally.

So far, we have seen many examples of heuristics to decide which nets to rip up and how to reroute them. Indeed, the choice of these strategies can have a huge influence on the effectiveness of a rip-up and reroute scheme at route completion, as well as on the time required for the route completion. Thus, this choice has a significant impact on the overall quality of a global routing tool. Commercial global routers often use sophisticated, highly tuned strategies for rip-up and reroute that yield route completion rates and runtimes superior to most other competing global routing approaches.

4.1.3 Hierarchical and Multilevel Routing

These approaches rely on the use of a hierarchy of routing grids. Thus, they are a natural extension of the traditional two stage decomposition of routing into global and detailed routing.

In most hierarchical schemes, the global routing is first carried out on a very coarse grid composed of "super-cells", and is then refined recursively by embedding the routes into successively finer routing grids, as illustrated in Fig. 4.4. The routing of all the nets on a coarse grid allows the generation of an approximate congestion map while still retaining some flexibility in the embedding of the nets on the final global routing grid. This helps overcome some of the problems inherent in net ordering, since the nets routed early still retain some flexibility of responding to a congestion map that captures all the nets, without requiring any rip-up and reroute. However, the routing of a net at a previous hierarchical level does constrain the embedding choices available to it at subsequent levels.

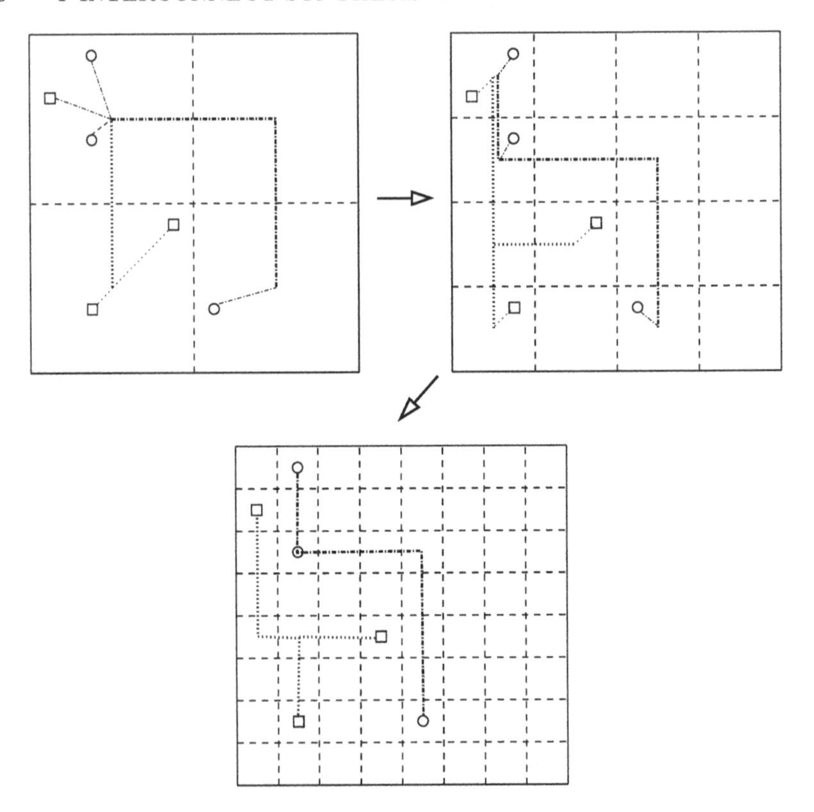

Fig. 4.4. Hierarchical global routing.

Another advantage of hierarchical schemes is that they help increase the capacity of the global router. Since the top level grid is much coarser than the final global routing grid, and super-cells at subsequent routing levels are not very dependent on each other (and can therefore be parallelized), a hierarchical global router can handle much larger designs than a corresponding flat router.

The first hierarchical global routing scheme was presented in [BP83]. The core of this algorithm is a scheme to refine the routing from one hierarchical level to another. In particular, [BP83] proposes two heuristics, based on divide-and-conquer and on dynamic programming respectively, to refine the routes in a row (column) of the original grid, represented as a $1 \times N$ ($N \times 1$) array of cells, to a $2 \times N$ ($N \times 2$) array of cells at the next routing level, as illustrated in Fig. 4.5. This work has been followed by many other variants for top-down hierarchical global routing over the years. These variants have relied upon maze routing, integer linear programming (as in [BP83] and [HL91]) or network flows (as in [CS98]) to carry out the routing at any given level of the hierarchy, proposing different schemes to refine the routes from one level of the hierarchy to the next.

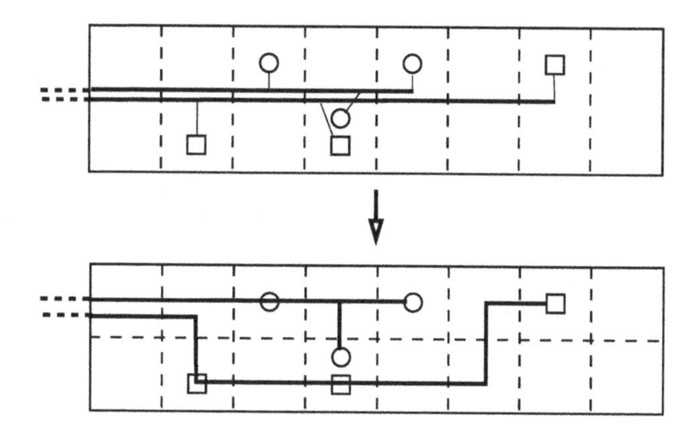

Fig. 4.5. Refining the routes in a row of super-cells across one routing level.

Since the routing decisions at the coarse levels of any top-down hierarchical global routing scheme are made with little visibility into the underlying global routing cells, these decisions can lock the routes into poor embeddings. This problem is avoided in bottom-up hierarchical schemes such as that proposed in [Mar84]. In this scheme, the routing grid is successively coarsened until the entire layout can be represented by a single super-cell. At any level of the hierarchy, only the nets that lie entirely within the super-cells of the current routing grid are routed. Therefore, when the routing grid is coarsened to the next level, many of the nets that previously crossed super-cell boundaries now become available for routing. Although this approach captures the local congestion well, it suffers from a lack of global visibility during the early routing decisions, potentially resulting in poor routes for the long global nets.

The problems inherent in pure top-down or bottom-up hierarchical schemes have encouraged researchers to develop several hybrid schemes such as [LHT90] and [HT95]. The approach in [LHT90] proposes parameter-controlled top-down refinement of the coarse grid, allowing some portions of the routing graph to become finer than others. In contrast, [HT95] embeds a top-down hierarchical loop inside a bottom-up hierarchical loop. All nets are first decomposed to two-pin subnets. At any level of the hierarchy, the only nets (or subnets) that are considered are those whose endpoints lie in super-cells lying adjacent to or diagonally across each other, with a shared super-cell boundary or corner. The routing of the nets at any given hierarchical level in the outer bottom-up loop is followed by a top-down rip-up and rerouting of all the nets that have already been handled at all the finer granularities of the routing grid, using a maze router. This approach usually yields lower routing congestion than pure top-down hierarchical routing or pure rip-up and reroute techniques.

In recent years, these hybrid approaches have been developed further into true multilevel routing. The fundamentals of multilevel optimization schemes will be discussed in Section 5.1.3 in Chapter 5 (since, so far, these techniques have been more effective in the context of placement than in the context of routing). The application of such schemes to multilevel routing, presented in [CFX+05] and [CL04], uses a "V-shaped" flow that first coarsens the routing grid in a bottom-up pass, and then refines it back to the granularity of the global routing cells in a top-down pass. The initial coarsening pass is used to estimate and reserve routing resources required locally for the current level of the hierarchy. At each new level of the hierarchy, this estimate includes the resources for nets lying at lower levels of the hierarchy as well as the nets newly exposed at the current level. However, in contrast to pure bottom-up approaches to hierarchical routing, the actual routing of these nets is deferred to the end of the coarsening pass. Once the number of super-cells falls below a certain threshold, the coarsening pass is terminated and all the nets are routed using a multicommodity flow based algorithm. These initial routings are then refined during a refinement pass of the routing grid using a modified maze search. In contrast to traditional top-down hierarchical schemes, coarser level routes do not act as hard constraints to finer level routes. Thus, finer level routes can deviate from their coarser counterparts when the more detailed information about the local congestion and resources warrants it. Because of this flexibility, multilevel global routing usually yields better quality layouts than hierarchical schemes.

4.1.4 Multicommodity Flow based Routing

A multicommodity flow problem involves the transportation of a given number (say, k) of commodities from their respective sources to their respective sinks in a given network. Each edge e in this network has a given capacity $c(e)$, which serves as an upper bound on its usage $u(e)$ that is the sum of the flows of all the commodities routed along that edge. In the context of global routing, the commodities are the nets N_i $(i = i, \ldots, k)$ that are to be routed. An advantage of multicommodity flow formulations for global routing is that they can provide quality assurances on their solution, in contrast to the other heuristics discussed earlier. In other words, they can be used to determine whether a feasible routing exists at all for a given routing problem, as well as to find a routing that is within a specified bound of the optimum routing (if it exists).

Although the global routing problem can be formulated as a multicommodity flow either in terms of edges or in terms of routing trees, it is the latter formulation that is more widely used. In this formulation, let $\mathcal{T}_i = \{T_{i,1}, \ldots, T_{i,j_i}\}$ refer to a list of candidate routing trees for net N_i. Furthermore, let the 0,1-variable $x_{i,j}$ $(i = 1, \ldots, k, j = 1, \ldots, j_i)$ be defined as 1 if and only if tree $T_{i,j}$ is selected for net N_i. Then, the multicommodity flow formulation of the global routing problem can be expressed as an integer linear program as follows:

$$\begin{aligned} \text{Minimize} \quad & \hat{\lambda} \\ \text{subject to} \quad & \textstyle\sum_{T_{i,j} \in T_i} x_{i,j} = 1, \quad i = 1, \ldots, k, \\ & \textstyle\sum_{i,j: e \in T_{i,j}} x_{i,j} \leq \hat{\lambda} c(e), \forall e \in E, \\ & x_{i,j} \in \{0,1\}, \quad i = 1, \ldots, k, j = 1, \ldots, j_i. \end{aligned} \tag{4.1}$$

This formulation is also referred to as the *concurrent multicommodity flow* formulation; it seeks to minimize the maximum flow in the edges of the network. In this formulation, the first family of constraints, also referred to as the *demand constraints*, states that (flows equivalent to) exactly one routing tree will be selected for each net. The second family of constraints is also referred to as the *bundle inequality*, and captures the intuition that the total usage of a routing edge for selected routing trees cannot exceed its capacity (by requiring that $\hat{\lambda}$ be at most one for a feasible routing). The third family of constraints (*i.e.*, the integer constraints) specifies that a routing tree cannot be "partially" selected; instead, it must either be used fully or not at all. Since the solution of an integer linear program is an NP-hard problem, it is usually difficult to solve large multicommodity flow problems optimally. Instead, a commonly used approach relies on relaxing the integer constraints so that $x_{i,j} \in [0,1]$, and then solving the resulting linear program. The resulting *fractional flow* is then rounded up to integer values of $x_{i,j}$ using various heuristics.

Alternative formulations for global routing using multicommodity flows associate a cost with each network edge, and then attempt to minimize the total cost of all the commodity flows in the network. One of the earliest multicommodity flow based approaches to global routing [SK87] is an example of this class of formulations; the rip-up and reroute scheme used in this approach was briefly discussed in Section 4.1.2.

The first multicommodity flow algorithm for global routing that had a theoretical bound from the optimal solution was presented in [CLC96]. It uses the concurrent flow formulation and is based on a two-terminal multicommodity fractional flow algorithm presented in [SM90]. In this approach, the integer constraints in Program (4.1) are relaxed, and the resulting linear program and its dual are solved. The theory of duality implies that a feasible solution to the dual program (specified below as Program (4.2)) is a lower bound to the optimal solution of the original program (4.1). This is exploited by the algorithm of [SM90] that iteratively pushes the solutions to these two programs closer. The final gap between these two solutions provides the theoretical bound on the quality of the solution. Finally, the fractional solution obtained by the relaxed linear program is rounded up to integer values using randomized rounding, with any remaining overflow resolved using rip-up and reroute heuristics.

The dual of Program (4.1) can be written as:

$$\begin{array}{lll} \text{Maximize} & \sum_{N_i} \theta_i & \\ \text{subject to} & \sum_e k(e)l(e) = 1, & \\ & \sum_{i,j:e \in T_{i,j}} l(e) \geq \theta_i, \ \forall N_i, \forall T_{i,j} \in \mathcal{T}_i, & \\ & l(e) \geq 0 \qquad \forall e. & \end{array} \tag{4.2}$$

In this formulation, the dual variable $l(e)$ is referred to as the weight of the edge e and initialized to $1/c(e)$, and $k(e)$ is the cost of pushing a unit flow through that edge. The variable θ_i represents the throughput of the flow for the net N_i.

The algorithm initially constructs Steiner trees for all the nets without considering the bundle constraints; this usually results in a large initial value of $\hat{\lambda}$. This value is iteratively decreased by reweighting the edges (so that congested edges have a higher weight) and reconstructing minimum weight Steiner trees.

More recently, a faster and simpler multicommodity flow algorithm was presented in [GK98] and applied to the global routing problem in [Alb00]. Unlike the algorithm in [CLC96] in which a fraction of the flow in a highly congested tree is switched to a less congested tree, this algorithm adds an incremental flow to a less congested tree in each iteration, without changing any of the previously placed flows. Finally, the flow on each edge is scaled back by the number of iterations. Just as with [CLC96], this approach also provides a theoretical bound on the quality of the solution.

Although multicommodity flow based approaches to global routing have shown good progress in recent years, it has been a challenge to scale them up to today's large global routing problems and yet obtain quality-versus-runtime tradeoffs that are comparable in quality to the best rip-up and reroute heuristics.

4.1.5 Routing using Simulated Annealing

Simulated annealing [KGV83] has been successfully applied to numerous computationally difficult combinatorial optimization problems, including several in the area of physical design (such as floorplanning and placement). It is a means of reducing the likelihood of getting trapped in local minima within a complex solution space. Given a candidate solution, potential moves leading to a new candidate solution are generated and evaluated. Any move that reduces the cost function is accepted. Furthermore, in contrast to greedy heuristics, even a move that increases the cost function has a finite probability of acceptance. This probability is given by $e^{-\Delta C/T}$, where ΔC is the increase in the cost and T is a parameter commonly referred to as the *temperature*. Thus, the probability of the acceptance of a move decreases exponentially with the extent of the increase in the cost. Furthermore, a reduction in the temperature also decreases the probability of the acceptance of a move that degrades the cost function. As the exploration of the solution space proceeds, the temperature is progressively reduced.

This approach was applied to the global routing problem quite early [VK83]. Although the original formulation of global routing using simulated annealing was restricted to two-pin nets and routings that had no more than two bends, it was extended to generic global routing as part of the TimberWolf layout system [SS86]. TimberWolf applies simulated annealing for both placement and routing. A global routing move in this framework consists of changing the routing of a net currently crossing some congested routing edge, from one candidate routing to another. The cost function to be minimized is the total routing overflow within the entire routing grid. Generating all possible routings (and topologies) for all the nets up front is not practical. Therefore, it is possible to generate new routings for a net on the fly. As is typical with most simulated annealing based algorithms, this approach to global routing can involve large runtimes. However, the longer the algorithm is run, the better is the quality of the resulting solution.

Other similar move-based approaches such as simulated evolution and genetic algorithms have also been used to develop formulations for global routing, although they have not yet been demonstrated as competitive against industrial tools.

4.1.6 Routing using Iterative Deletion

Iterative deletion, presented in [CP92], is a technique that inverts the traditional constructive paradigm of global routing in which one route after another is added to the layout. Instead, iterative deletion conceptually begins with all possible routings for all the nets, and then iteratively deletes the most expensive Steiner segments from the routings, provided that they do not cause the net they belong to, to become disconnected. This is continued until the route for each of the nets has been reduced to a tree, with no redundant Steiner segments. The generation of all possible routings up front enables the early approximation of the final congestion map, theoretically allowing accurate congestion costs to drive the selection of the Steiner segments to be deleted. This helps ameliorate the net ordering problem to some extent. However, if the number of alternative routings for each net is large, the congestion map is no longer accurate.

Of course, constructing all possible routings for all the nets is prohibitively expensive in practice. Therefore, impractical routings (such as those that involve significant detours) are not considered. Even for a single net, given n Steiner points, iterative deletion conceptually reduces the $O(n^2)$ Steiner segments possible for its routing down to some $n-1$ segments forming a tree. However, even the $O(n^2)$ iterations required for this process can be impractical. Therefore, heuristics are used to prune down the number of Steiner segments to $O(n)$ by not generating segments that are unlikely to be retained.

Although this pruning improves the runtime as well as the accuracy of the initial congestion map to some extent, it can have a significant impact on route completion in congested designs (since detoured routes that may be required in such designs are not a part of the pruned set of routing choices for iterative deletion). As a result, iterative deletion has not found widespread acceptance in industrial global routing tools.

4.2 Congestion Management during Detailed Routing

Whereas the role of global routing is to perform route planning in a way that the congestion due to global routes is spread uniformly across the chip and the routing overflows in global routing bins are minimized, the responsibility of legalizing the global routing and resolving all remaining routability issues lies with detailed routing. In particular, detailed routing is the primary vehicle for resolving local pin accessibility issues, as well as eliminating all the routing overflows that remain after global routing. Furthermore, in contrast to global routing that works with a simplified routing model, detailed routing must comprehend all the layout rules specified by the process technology. As manufacturing technologies descend yet deeper into the sub-wavelength realm, these layout rules become exponentially more complicated, with more and more context-dependent and non-local rules being specified, leading to the process of detailed routing becoming yet more compute-intensive. However, given the local scope of detailed routing, the primary objective for a detailed router is legal route completion even in performance-driven designs, in contrast to global routers whose behavior can differ significantly depending on the delay, signal integrity and other constraints specified for the design.

Consequently, in contrast to global routing that operates upon the entire chip or design block, detailed routing works on a very local three-dimensional switchbox. It usually starts with a switchbox that is approximately the same size as a handful of global routing bins. If it is not able to legalize the routing within that switchbox, it then expands it to encompass parts of the surrounding bins also, in order to have a slightly larger scope for optimization. However, the compute-intensive nature of detailed routing implies that the switchboxes that it operates on can never become much larger than a few dozen global routing bins. On the other hand, this local scope of detailed routing means that it is easily parallelizable, in contrast to global routing. Another heuristic that is sometimes helpful in cases of extremely localized, pin accessibility caused hot spots is that of shrinking the corresponding switchboxes to a small neighborhood of the hot spots, and then using more exhaustive switchbox routing techniques on these shrunk switchboxes.

In order to legalize a switchbox, the detailed router first creates a "virtual pin" at each boundary crossing where some route enters or leaves the

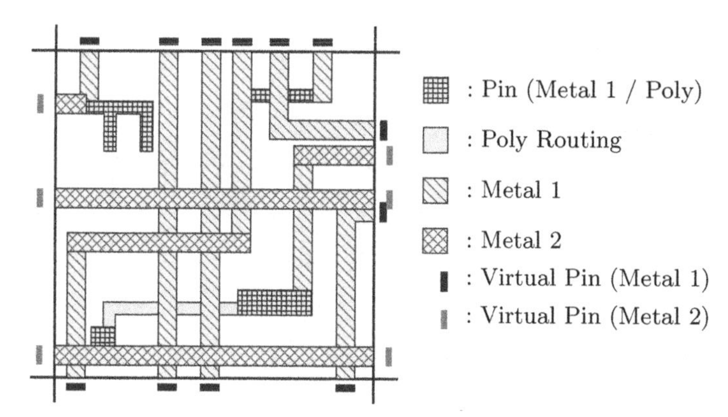

Fig. 4.6. A simplified, two metal layer switchbox for detailed routing (with vias and contacts omitted for simplicity), showing virtual pins created at the boundary of the switchbox.

switchbox, as shown in the picture of a simplified two metal layer switchbox illustrated in Fig. 4.6. Then, fixing these virtual pins at their locations, it attempts to rip up and reroute all or part of the routing within the switchbox in order to legalize it using various heuristics. However, the fixed locations of the virtual pins can be a huge handicap to the successful operation of the detailed router. Their adverse impact is minimized by creating overlapping switchboxes, as shown in Fig. 4.7. Iterating over these overlapping switchboxes allows the virtual pins themselves to be easily ripped up, thus reducing the adverse impact of the "boundary effect" due to poor initial virtual pin positioning, as shown in Fig. 4.8. In the example depicted in this figure, the overlapping of adjacent switchboxes allows the removal of the unnecessary wire detour created to access the virtual pin at the original switchbox boundary.

Detailed routers also freely use non-preferred direction routing on different metal layers (especially while accessing pins on the lower metal layers), if it helps complete the routing inside the switchbox[3]. The creation of expanded, overlapping switchboxes can also help resolve incidents of local routing overflows, by pushing some routes from an overflowing bin lying on the boundary of a congested region into an adjacent uncongested bin.

[3] The use of a non-preferred routing segment connected to a routing segment on the same metal layer that is routed in the regular direction can help avoid the use of a via or contact, thus easing the congestion inside the switchbox. In contrast, global routers rarely use non-preferred direction routing because it creates a blockage across multiple tracks within that routing layer, that can potentially impact the porosity of that layer for global routing significantly. This is a much smaller problem for the detailed router since the scope of the router is limited to a switchbox. Consequently, the routability impact of the non-preferred direction route within the switchbox can be easily and accurately estimated.

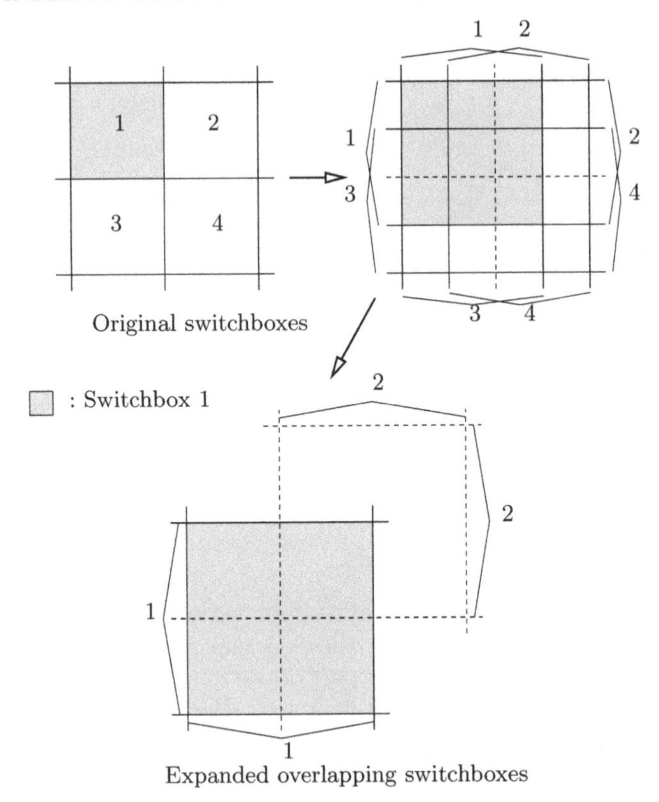

Fig. 4.7. Creation of overlapping and expanded switchboxes.

Different detailed routers available commercially differ in the strategies they use to decide how and when to expand the switchboxes. In general, the first pass of detailed routing goes over all the initial switchboxes (comprising of small rectangular clusters of global routing bins covering the entire routing area with little or no overlaps at their boundaries). In the next iteration, it may try to fix more routability problems by creating larger overlaps among the switchboxes. Subsequent iterations usually focus on the congestion hot spots, employing various strategies for switchbox expansion, contraction or overlap to try and complete legal routes in these regions. The selection of the strategies for this switchbox manipulation can have a significant impact on the effectiveness of the detailed router at legal route completion.

Various techniques have been proposed to solve the routing problem within each switchbox taking into account the geometric design rule constraints. Although the most effective techniques in industrial tools rely on various rip-up and reroute heuristics, several other formulations have also been proposed in the literature on this subject. These include mathematical programming formulations, multicommodity flow based formulations, as well as formula-

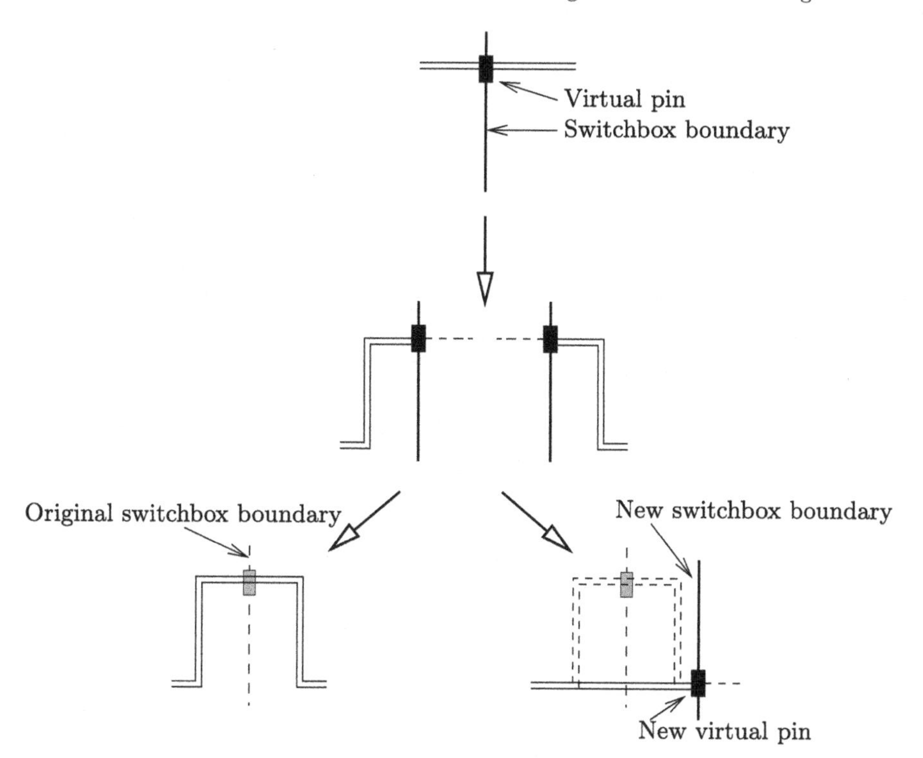

Fig. 4.8. Overlapping switchboxes can help reduce the adverse impact of poor initial virtual pin positioning.

tions that use move-based heuristics such as simulated annealing and genetic algorithms.

4.3 Congestion-aware Buffering

Since wires scale much worse than devices, an optimized interconnect often requires additional buffers when shrunk to the next process technology node[4]. Indeed, recent technology studies [Con97, SMC+04] have studied and quantified the rapid increase in the number and fraction of buffers in designs as they scale across process nodes. As the number of buffers in a design block increases, it begins to have a significant impact on the routability of the nets within the block. As mentioned earlier in this chapter and illustrated in Fig. 4.9, this is because of three reasons. Firstly, every inserted buffer implies two additional via stacks to access the buffer from the layer that the net is routed on; these via stacks use up valuable routing resources, especially on the lower layers that

[4] Under first-order scaling assumptions, optimal inter-buffer distances shrink at $0.586\times$ per generation, in contrast to the normal geometric shrink factor of $0.7\times$.

are crucial to ensuring pin accessibility during detailed routing. Secondly, if a buffer cannot be legally placed at its desired location, it may force a routing detour (as well as additional vias), adding to the routing congestion. Finally, since the placement of a buffer is equivalent to fixing the location of a Steiner point in the routing of a net, inserting a buffer into a net and placing it cuts down the router's flexibility in ripping up and rerouting this net at a subsequent stage in response to the evolving local congestion profile. Given the important role that rip-up and reroute plays in route completion, this loss of flexibility due to buffering can degrade the routability of a design significantly. Therefore, it is important to carry out the buffering of nets in a holistic fashion that understands the impact of the buffers on the routability of the design.

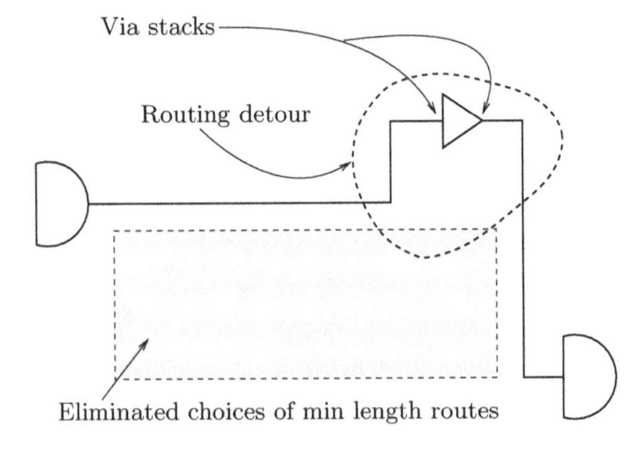

Fig. 4.9. The congestion impact of inserting a buffer into a net.

Many of today's design methodologies use buffer blocks to provide the buffering required by long nets. Furthermore, numerous algorithms have been proposed for the synthesis of individual buffered nets. Several of these algorithms can handle routing blockages, buffer insertion blockages, predefined buffer locations, and routing congestion cost. However, the study of their effect on the overall routability of the design is a recent phenomenon. In this section, we will discuss the routability-aware design of buffer blocks, as well as the design of a more fine-grained buffering algorithm for nets that tries to tradeoff between environmental considerations (such as blockages, congestion and cell density) and performance considerations, without sacrificing runtime (in order to make it scalable to a large number of buffered nets).

4.3.1 Routability-aware Buffer Block Planning

With a significant fraction of global nets requiring buffering, several design methodologies in use today rely on buffer blocks to supply these buffers. These blocks are often fitted into the channels and other dead space between the circuit blocks during the floorplanning stage (as illustrated in Fig 4.10), and are then utilized during the global routing stage. Such methodologies are especially appropriate for large, complex designs such as microprocessors that are designed hierarchically (*i.e.*, the circuit blocks are treated as impermeable black boxes during the top-level chip assembly) or designs that involve a large number of third-party cores that are provided as hard macros with immutable layouts.

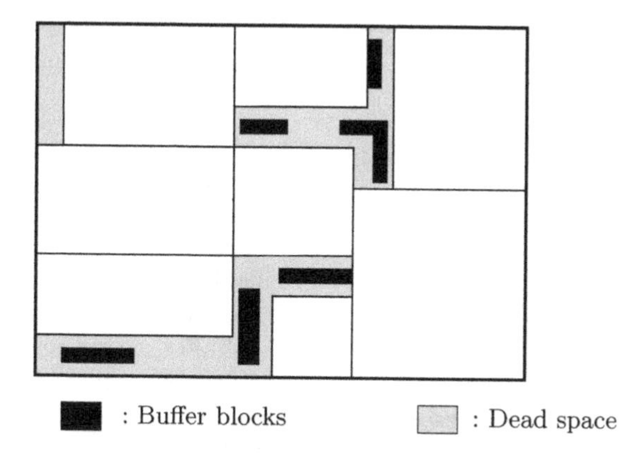

Fig. 4.10. Buffer blocks placed in routing channels and other dead space within the floorplan of a chip.

The delay of a long buffered wire does not change appreciably under small perturbations of its buffers from their optimal positions (especially if they can be subsequently sized appropriately). This is because the signal propagation speed versus the inter-buffer distance curve for a long buffered wire is relatively flat around the optimum inter-buffer separation L_{buf}^{opt}, as illustrated in Fig. 4.11, so that a small permissible delay degradation corresponding to Δd in the figure can result in a considerable layout flexibility Δl in the positioning of the buffers. The curve for the cross-coupled noise is also similar in shape. This observation, long used by circuit designers to create layout flexibility, has been used in [CKP99] to introduce the concept of a *feasible region* for each buffer required for a net. A buffer placed within its feasible region is guaranteed to allow the net to meet its target delay, provided that the remaining buffers (if any) on the net are placed optimally with respect to this buffer. For a net that is not very timing-critical, the feasible region for each of its buffers can be quite large, so that the likelihood of its overlapping with at least one of the preplaced buffer blocks is quite high. Thus, each buffer required by a global net in the design can be placed within a buffer block that overlaps with its feasible region, without perturbing the circuit blocks within the design.

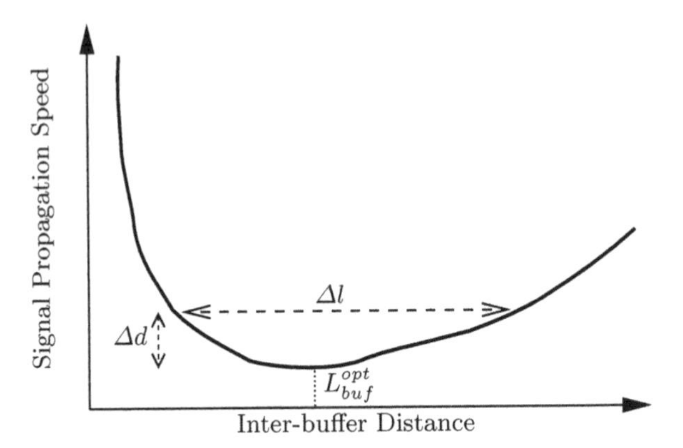

Fig. 4.11. The variation of the signal propagation speed of a long uniformly buffered wire as its inter-buffer distance is varied, for some given metal layer.

There are three problems with the feasible region approach described in [CKP99]. Firstly, the feasible regions for the buffers on a single net are not independent, so that the positioning of a buffer near the boundary of its feasible region can substantially shrink the feasible regions for the remaining buffers on that net. Secondly, the buffer blocks can act as hot spots for routing congestion. Finally, a wire crossing a large block could have no buffers available to it unless it detours significantly to the nearest buffer block, resulting in degraded delay and wirelength for the net and increased congestion.

The first two of these problems have been addressed in a work on routability-driven buffer block planning in [SK01] that enhances the feasible region concept to that of *independent feasible regions* (IFRs). (In addition, the third problem mentioned above is addressed in the fine-grained buffer insertion approaches described in Section 4.3.2). The IFR of a buffer is the region within which it can be placed while meeting the delay constraint on the net, assuming that the remaining buffers required for that net are placed anywhere within their respective IFRs (in contrast to the requirement of their being placed optimally with respect to this buffer, that was used to define feasible regions). Note that the IFR of a buffer on a net that requires a single buffer is the same as its feasible region. Furthermore, the IFRs of the buffers on a net with the fewest possible buffers must be disjoint (else, placing a buffer within the intersection of two IFRs will reduce the buffer count of the net).

Assume that a two-pin net of length L when routed without any bends on a particular metal layer with resistance per unit length given by r and capacitance per unit length given by c requires n buffers. Then, assuming that the route of the net spans the interval $(0, L)$, the IFR for its ith buffer is an interval given by:

$$IFR_i = (x_i^* - W_{IFR}/2, x_i^* + W_{IFR}/2) \cap (0, L),$$

where x_i^* is the optimal location of the ith buffer and W_{IFR} is the width of each IFR. If the output resistance of the driver and a buffer is R_d and R_b respectively, and the input capacitance of a buffer and the sink is C_b and C_s respectively, then, under the assumption of the Elmore delay model [Elm48, RPH83], the location x_i^* of the ith buffer along the net is given by [AD97]:

$$x_i^* = (i-1)v_L^* + u_L^*$$

for $i \in \{1, \ldots, n\}$, where:

$$u_L^* = \frac{1}{n+1}(L + \frac{n(R_b - R_d)}{r} + \frac{(C_s - C_b)}{c}),$$

and,

$$v_L^* = \frac{1}{n+1}(L - \frac{(R_b - R_d)}{r} + \frac{(C_s - C_b)}{c}).$$

With these buffer locations, the optimal Elmore delay D_{opt} for the buffered net is given by:

$$D_{opt} = D(R_d, C_b, x_1^*) + D(R_b, C_s, L - x_n^*) + \sum_{i=1}^{n-1} D(R_b, C_b, x_{i+1}^* - x_i^*) + nT_b,$$

where T_b is the intrinsic buffer delay. The term $D(R, C, l)$ used in this expression stands for the usual Elmore delay of a wire segment of length l with sink load C being driven by a gate with output resistance R that is being modeled using the π-model, and is given by:

$$D(R, C, l) = R(cl + C) + rl(\frac{cl}{2} + C) = \frac{rcl^2}{2} + (Rc + rC)l + RC.$$

Then, if the delay budget for the net is D_{tgt} (with $D_{tgt} \geq D_{opt}$), the width of the (one-dimensional) IFR is given by:

$$W_{IFR} = 2 \cdot \sqrt{\frac{D_{tgt} - D_{opt}}{rc(2n-1)}}.$$

For a buffer on a two-pin net that is routed with bends, its two-dimensional IFR is defined as the union of all the one-dimensional IFRs (computed as described above) of that buffer on all the monotonic (*i.e.*, minimum length) Manhattan routes between the source and sink of the net. Therefore, two-dimensional IFRs are convex octilinear polygons with boundaries that are horizontal, vertical or diagonal (*i.e.*, at $\pm 45°$).

The route monotonicity requirement introduces a certain degree of dependence between the IFRs of the different buffers inserted in a net whose routing requires at least one bend. Consider the example in Fig. 4.12, that shows a net requiring two buffers whose IFRs are shown as IFR_1 and IFR_2. If the selected buffer locations are b_1 and b_2, they introduce a non-monotonicity within the routing as shown in the figure, even though both locations lie within their respective IFRs. Therefore, their insertion cannot guarantee that the target delay D_{tgt} will be met by the net. In order to avoid this problem, the IFRs of buffers that have not yet been inserted within a net must be adjusted in order to eliminate subregions resulting in non-monotonic routing, every time a location is selected (or narrowed down) for a buffer on the net. Thus, for the example in Fig. 4.12, the assignment of the first buffer to the location b_1 will result in the elimination of the shaded region of IFR_2 in order to guarantee a monotonic routing.

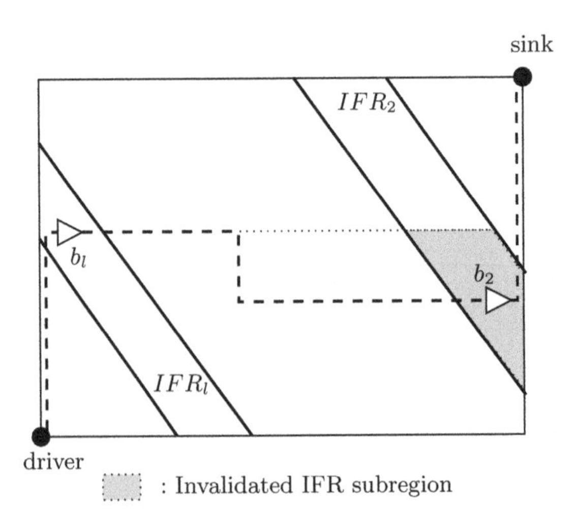

Fig. 4.12. The dependence between the IFRs of the buffers to be inserted in a net, caused due to non-monotonic routing, that results in the invalidation of location b_2.

In order to accomplish a routability-driven assignment of buffers to specific locations within their IFRs, a two-level tiling is constructed on the floorplan as shown in Fig. 4.13, with the coarser level corresponding to the granularity used for global route planning, and the finer level corresponding to specific sets of buffer locations within each IFR. Note that the finer tiling is not required outside of the IFRs of the buffers. Each fine-level tile is referred to as a *candidate buffer block* (CBB). Now, a bipartite graph $\mathcal{G}$ can be constructed to represent the set of all possible buffer assignments for all the buffers on all the global nets that require buffering. If $\mathbf{B}$ is the set of all the buffers on all the nets that must be assigned to specific CBBs, and S_b is the set of all the CBBs, then the edge set $E(\mathcal{G})$ of $\mathcal{G}$ is given by:

$$E(\mathcal{G}) = \{(b, c) : b \in \mathbf{B}, c \in S_b\}.$$

Each edge in this bipartite graph is costed using a combination of the corresponding expected routing and buffer congestions. In particular, the im-

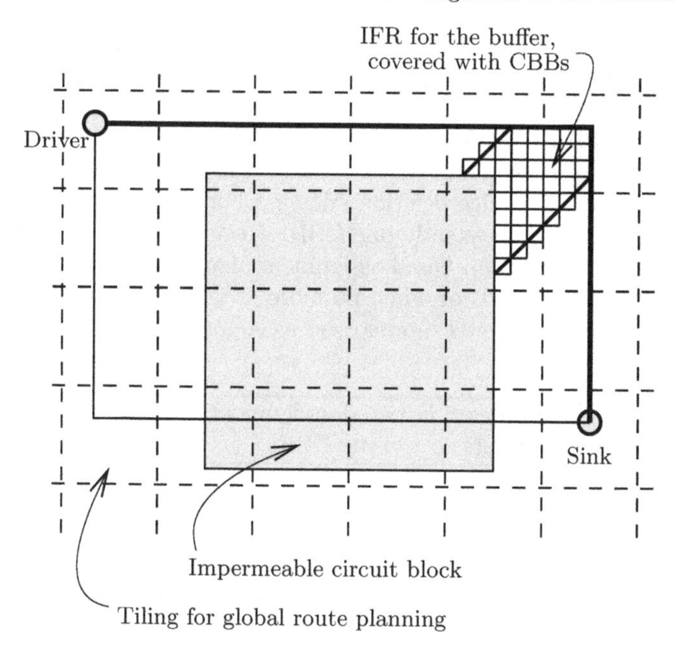

Fig. 4.13. The two-level tiling of the floorplan in order to create CBBs, illustrated for a single net requiring a single buffer.

plementation in [SK01] sets the cost of an edge $e = (b, c)$ to $CC(e)^{p_1} \cdot BB(e)^{p_2}$, where $CC(e)$ is the routing congestion cost of the buffer b and $BB(e)$ is the buffer congestion cost of the CBB c, and p_1 and p_2 are positive coefficients summing up to 1 that allow a tradeoff between the two congestion costs. The routing congestion cost of a buffer b is the maximum normalized congestion cost (*i.e.*, a supra-linear function of the ratio of the expected number of routes passing through a routing bin to the routing capacity of that bin) among all the routing bins lying along the one-bend routings of the two subnets attached to b, assuming that b is to be placed at the current CBB c (and the other terminal of the attached net is placed at the centroid of all its feasible CBBs, in case it too is a buffer that has not yet been assigned to a CBB). The supra-linear nature of the normalized congestion cost severely penalizes routings that pass through areas of heavy congestion. The buffer congestion cost of a CBB is infinite if all the buffers within it have already been allocated. Otherwise, it is defined as $1/\min\{B_c, B_{\max}\}$, where B_c is the number of IFRs overlapping this CBB, and $B_{\max}$ is the maximum number of buffers that can be placed within this CBB.

Once the graph $\mathcal{G}$ has been set up, buffer assignment is done by iteratively deleting the most expensive redundant edge from $\mathcal{G}$ (*i.e.*, an edge whose removal will not leave a buffer node unconnected). After each edge deletion (which corresponds to the elimination of some buffer location candidates for some buffer that lies on, say, net N), the IFRs for the remaining unassigned buffers on N are updated to eliminate any CBBs whose selection would result in non-monotonic routes. A fast scheme for this monotonicity update is presented in [SK01]. The deletion of an edge can also affect the congestion cost and the buffer cost of the edges remaining in the graph; these costs are also updated after each edge deletion. When a node in $\mathcal{G}$ corresponding to a buffer is left connected to exactly one CBB, it is considered assigned to that CBB. This iterative deletion based assignment algorithm terminates when all buffers have been assigned (or when no more assignments are possible). The operation of this algorithm is summarized as Algorithm 5.

Algorithm 5 Iterative deletion based assignment algorithm for routability-driven assignment of buffers to specific CBBs

1: Compute the IFR for each buffer $b \in \mathbf{B}$
2: Build a fine-grained tiling structure of CBBs over the IFRs for the buffers in $\mathbf{B}$
3: Obtain CBB set S_b for each $b \in \mathbf{B}$
4: Generate the bipartite graph $\mathcal{G}$ and cost all its edges
5: **while** there exists an unassigned buffer, and further edge deletion is possible, **do**
6: Delete the highest cost redundant edge e of $\mathcal{G}$
7: Update all affected IFRs for monotonicity, deleting the invalidated edges from $\mathcal{G}$
8: Update the congestion matrix
9: Update edge costs
10: **if** buffer b' has only one edge (say, to CBB c'), **then**
11: Assign buffer b' to CBB c'
12: **end if**
13: **end while**

Feasible regions and IFRs are defined with respect to two-pin nets that are routed monotonically. Thus, a multipin net must be broken down into two-pin subnets and its delay budget must be distributed among these subnets, before the IFR concept can be used for buffer insertion using buffer blocks in a way that meets the delay budgets for the subnets.

4.3.2 Holistic Buffered Tree Synthesis within a Physical Layout Environment

Although buffer blocks that are fitted into channels and other dead space between design blocks are commonly used today to provide the buffering for long nets, this methodology has limitations that make it difficult to scale to future designs that will require considerably larger numbers of buffers. With the rapidly decreasing inter-buffer distances in upcoming process technology nodes, the utility of buffer blocks decreases as the detours required by nets in order to access these blocks become an increasingly large fraction of the desired inter-buffer distance itself, leading to significant performance degradation[5]. Additionally, since all the wires requiring buffering must access these blocks, there is considerable contention for routing resources in their vicinity, leading to heavy congestion and potential unroutability. Buffer blocks also tend to become thermal and power grid voltage droop hot spots. Consequently, recent research has focused on finer-grained buffer insertion approaches that try to distribute buffers to their ideal locations along the nets requiring them.

One of the first works to present a methodology for integrated buffer and wire planning on a given placement that allowed the buffers to be placed anywhere in the design depending on the local cell density was [AHS+03]. The global routing and buffering framework proposed in that work is as follows:

1. Use a fast, performance-driven heuristic (such as the C-TREE algorithm from [AGH+02] for multipin nets and length-based buffering for two-pin nets) to construct congestion-oblivious Steiner trees for all the nets.
2. Rip-up and reroute the nets iteratively to reduce the routing congestion.
3. For all the nets that require buffers, perform a fast buffer insertion driven by dynamic buffer congestion costs.
4. Rip-up, reroute, and reinsert buffers on the nets to reduce both wire and buffer congestion.

One of the major contributions of this work is the notion of the buffer congestion cost of a global routing bin. For each bin, it compares the sum of the number of buffers already assigned to that bin and the number of buffers expected to be required for the currently unrouted nets that may pass through that bin, with the maximum number of buffers that can be accommodated by that bin, to determine a buffer congestion cost analogous to the traditional routing congestion cost used by global routers. This allows the buffers to be embedded along the routings of the nets in bins that can accommodate them, thus distributing them across the design without causing any large routing detours.

This framework has served as the basis for several major improvements such as [AGH+04], [SHA04] and [AHH+04]. One of the significant advances, proposed in [AGH+04], allows the Steiner points of the tree embeddings obtained after the routing congestion driven rip-up and reroute step to be moved locally in response to the buffer congestion cost, as illustrated in Fig. 4.14 for the Steiner points s'_1 and s'_2, thus providing a fast feedback loop from the buffer insertion step to the earlier topology embedding step. However, since this is achieved by propagating multiple candidate solutions for each node

[5] Furthermore, the signal propagation speed versus inter-buffer distance curve illustrated in Fig. 4.11 is also becoming less "shallow" with process scaling, even in a normalized sense [Sax06], so that the performance degradation for the same detour (normalized to the optimal inter-buffer separation for the relevant metal layer and process generation) required to access a buffer block is worsening with each process generation.

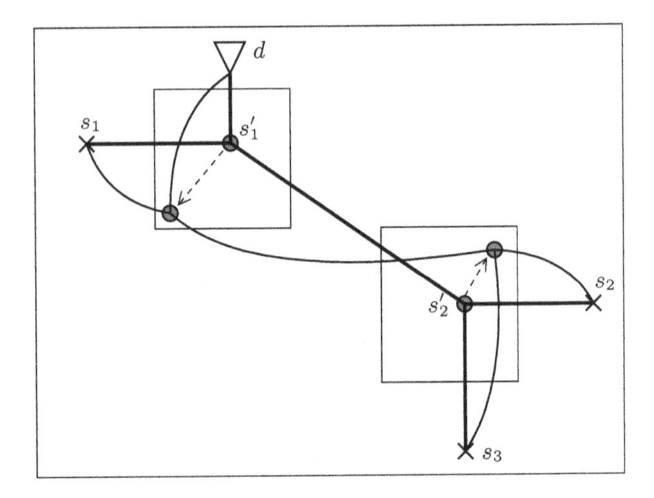

Fig. 4.14. Local buffer congestion driven perturbation of Steiner points of the embedding of a Steiner tree topology.

in a dynamic programming framework, the algorithm is not scalable to a large number of nets without sacrificing tiling granularity (and hence, performance). Furthermore, it does not distinguish between critical and non-critical nets while deciding the extent of perturbation allowed on their tree embeddings in response to local congestion costs. A more effective algorithm for integrated global routing and buffering under performance constraints that is targeted for use on a large number of nets is put together in [AHH+04] and is described next.

This algorithm, called BEN, is motivated by the observation that buffer insertion in non-critical nets can be considerably more flexible than in critical nets. Thus, regions with high buffer congestion (for instance, narrow channels between large, impermeable blocks, the holes within such blocks, and the boundaries of such blocks, or regions with high cell density, as illustrated in Fig. 4.15) as well as regions with high routing congestion should be used preferentially for the critical nets. This can be achieved by (re-)locating the Steiner nodes for the embedding of a net appropriately depending on the criticality of the net and the local buffer and routing congestion costs.

The basic framework for BEN is somewhat similar to that for the algorithm in [AHS+03] that was briefly discussed earlier in this section. Specifically, for any given placement, once a background cell density and routing congestion map has been generated, the synthesis of each multipin interconnect tree involves three main steps:

1. Use a fast, performance-driven heuristic (such as the C-TREE algorithm from [AGH+02]) to construct a congestion-oblivious Steiner tree for the net.

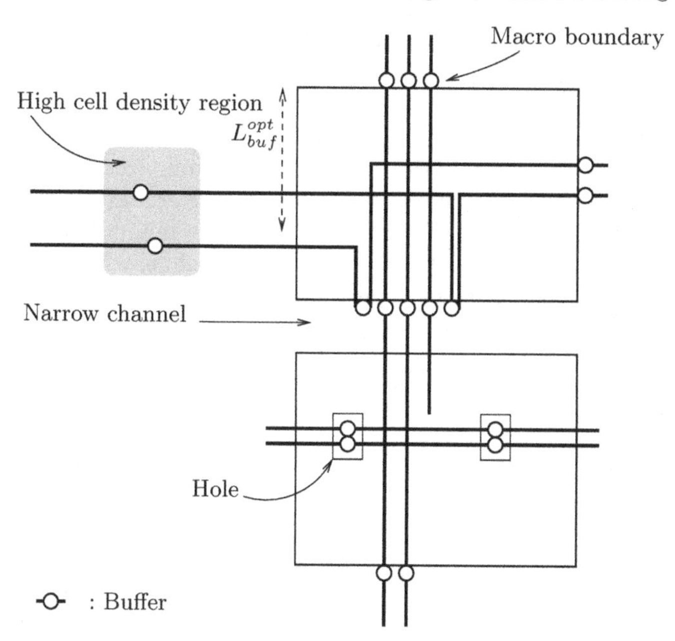

Fig. 4.15. High buffer congestion cost regions created in the vicinity of impermeable macro blocks or high cell density regions (L_{buf}^{opt} is the optimal inter-buffer distance).

2. Reroute the tree to preserve its topology even as its embedding is perturbed while navigating environmental constraints such as buffer congestion and routing congestion, keeping the criticality of the net in mind.
3. Remove the buffers inserted in step 2, and then reinsert and size them using a resource-aware variant (such as [LCL96]) of the dynamic programming based van Ginneken algorithm [Van90] using accurate, higher-order delay models.

The BEN algorithm targets Step 2 of the above framework. Since it is required to be fast in order to efficiently explore various embeddings of the tree topology, it cannot use expensive delay models or propagate multiple candidate solutions as in Step 3 (which is carried out on a topology embedding that is immutable). The primary innovation of BEN is a generalized costing mechanism that can handle both critical and non-critical nets, increasing the environmental awareness for the non-critical nets. This allows a traditional maze router to be run for each edge of the Steiner tree to the neighborhoods of the Steiner nodes that form its endpoints. The cost of routing a tree edge through a global routing bin b is expressed as:

$$cost(b) = 1 + K \cdot e(b) + (1 - K) \cdot TDC_{delay}(B(b)) + TDC_{drc}(B(b)), \quad (4.3)$$

where K (with $0 \leq K \leq 1$) is a parameter capturing the criticality of the net (with K being close to one for non-critical nets), $e(b)$ is an environmental cost modeling the routing and buffer congestions of the bin b, and the two $TDC(B(b))$ terms are costs modeling the maximum slew or load constraints (if present) and the quadratic delay of long unbuffered wires (whose computation will be discussed later). The parameter K can be reduced to some small value such as 0.1 for the most critical nets[6], which minimizes the impact of the environmental costs $e(b)$ for the bins used by the routes for these nets. On the other hand, a value of K close to one for the non-critical nets prevents them from using routing or buffer resources in regions where these resources are scarce (as in the examples depicted in Fig. 4.15). The environmental cost $e(b)$ for a bin b is defined as:

$$e(b) = \alpha d(b)^2 + (1 - \alpha) r(b)^2,$$

where $d(b)$ is the cell density of the bin (defined as the ratio of the total area of the cells placed within b to the area of b itself), $r(b)$ is the routing congestion of b (defined as usual as the ratio of the number of used horizontal or vertical tracks passing through b to the total number of corresponding tracks lying over b), and α $(0 \leq \alpha \leq 1)$ is a parameter that allows these two congestion metrics to be traded off against each other. While computing $e(b)$ in order to route a wire through the bin b, the appropriate value of $r(b)$ corresponding to the horizontal or vertical routing congestion is selected, depending on the direction in which the wire is being routed. The congestion terms are squared in the expression for the environmental cost in order to increase the penalty as either of these congestion terms approaches full utilization (*i.e.*, a value of one). These congestion terms can be raised to even higher powers to make the penalty for high utilization even more severe.

The first term (namely, one) in Equation (4.3) corresponds to the delay contribution of bin b (under an assumption of linear length-based delay). This assumption of linear delay is usually valid for nets that will be optimally buffered and that lie in regions having no blockages or high cell density regions (so that each inserted buffer can be placed optimally). On the other hand, large blockages can force long unbuffered stretches on the net, where the delay grows quadratically. In other words, the delay cost per bin grows linearly (instead of remaining constant) along these stretches. This intuition is captured by the third (TDC_{delay}) term in Equation (4.3). $TDC_{delay}(x)$ is a function whose definition is shown pictorially in Fig. 4.16. In this figure, if the number of consecutively blocked bins in which no buffer can be inserted is smaller than a threshold L_{delay}, the delay of the net can be assumed to be linear, so that the delay cost per bin is merely one. However, if the number of such bins exceeds L_{delay}, the linear cost per bin makes the overall delay of the portion of the net routed through these bins grow quadratically. The threshold L_{delay}

[6] Reducing K to zero causes the environmental costs to be completely ignored, which can result in wires passing through extremely congested regions as well as in illegal buffer placements, whose legalization can degrade the performance significantly.

is usually the longest distance over which linear delay growth can be obtained given the buffer library at hand. Note that this cost is relevant primarily to critical nets, hence the $(1 - K)$ multiplier associated with this term. This multiplier reduces the impact of the delay penalty of long, unbuffered sections of the routes for nets that are not timing-critical.

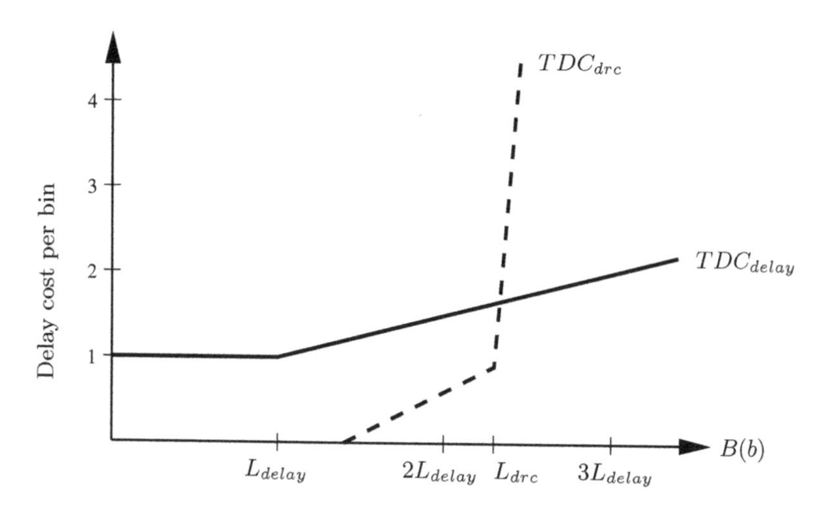

Fig. 4.16. Computation of $TDC(x)$ to model the delay cost due to bins blocked to buffers (as shown by the solid line) or design rule constraints for maximum load or slew (as shown by the dashed line).

The TDC_{delay} term can be easily computed by using a variable $B(b)$ at each bin b to keep track of the number of consecutive bins within which no buffer can be inserted, that has been seen by a route passing through b. Thus, if a route reaches a bin b' where no buffer insertion is possible from a bin b, $B(b')$ is set to $B(b) + 1$; on the other hand, if a buffer can be placed inside b', $B(b')$ is reset to 0. The only complication arises at branching points in the net; if T_l and T_r are, respectively, the number of blocked bins seen by the left and right children of a Steiner branch point located within the bin b, using $B(b) = \max\{T_l, T_r\}$ ignores the delay cost and load of one of the two children, whereas using $B(b) = T_l + T_r$ overestimates the delay impact. A reasonable compromise is using $B(b) = \sqrt{T_l^2 + T_r^2}$ in this case.

If slew and load constraints are present, they too can be captured by the same mechanism of $B(b)$, since they can be translated to length constraints using either empirical or analytical methods[7]. In this case, the TDC_{drc} function will grow to infinity for values of $B(b)$ that correspond to forbidden slews or loads that will cause a design rule violation (as shown by the dashed line

[7] Although [AHH+04] does not explicitly mention a TDC_{drc} term, the discussion of this term here is extrapolated from the ideas proposed in that work.

in Fig. 4.16). Furthermore, it is effective to have a region of supra-linear (say, quadratic) cost in the region immediately preceding the forbidden region in order to discourage routes whose inter-buffer separations are very close to violating these design rule constraints. (This creates a margin against the approximations inherent in translating the load and slew constraints into length constraints, as well as against subsequent design perturbations due to incremental changes). On the other hand, the cost of this term for bins with values of $B(b)$ much smaller than the length constraint L_{drc} is zero. Note that this cost applies to all nets, irrespective of their criticality.

In addition to accumulating the costs of the bins that its routing runs through, a timing-critical net also sees a cost because of its sinks, since they may differ in criticality. The cost of a sink s is initialized using:

$$cost(s) = (K - 1) \cdot RAT(s)/DpT,$$

where $RAT(s)$ (< 0) is the required arrival time at sink s, and DpT is the average delay of one bin in an optimally buffered two-pin net (*i.e.*, a net with linear delay). Thus, the more timing-critical sinks get a higher cost under this metric, encouraging shorter paths to those sinks from the driver. The multiplier of $(K - 1)$ causes this sink cost to be largely ignored for nets that have no timing-critical sinks. Thus, the accumulated cost at any node represents primarily the environmental costs for non-critical nets and the delay costs for the timing-critical nets. Consequently, the propagation of the costs of the two children when two branches are merged at a Steiner node should also be done differently. For a timing-critical net, the worst delay should be propagated upstream, whereas the two branch costs should be added up for a non-critical net. This is achieved by using:

$$cost(b) = \max\{cost(L), cost(R)\} + K \cdot \min\{cost(L), cost(R)\}$$

as the merging function at a Steiner node located within bin b and having children L and R.

This algorithm is sensitive to the accurate characterization of nets as critical or non-critical (by choosing an appropriate value of K for each net). Although the optimal selection of K for a net prior to its layout is very difficult (since the criticality of nets changes due to their delays after interconnect synthesis and global routing usually being different from their pre-layout predicted delays), this characterization problem can be largely avoided by first treating all nets as non-critical while laying them out with minimal use of the routing resources in congested areas, and then resynthesizing the nets that show up as timing-critical in this process. The suggested flow is to first generate congestion maps for buffer and routing congestion, and then lay out all the nets that require buffering using $K = 1$ (that will ensure very restricted use of buffers and routing tracks in congested regions). Next, nets that still have negative slacks can be ripped up and resynthesized with a slightly smaller value of K (such as $K = 0.7$). This process can be repeated for nets that still have a negative slack after the latest iteration with a yet smaller value of K, all the way down to, say, $K = 0.1$.

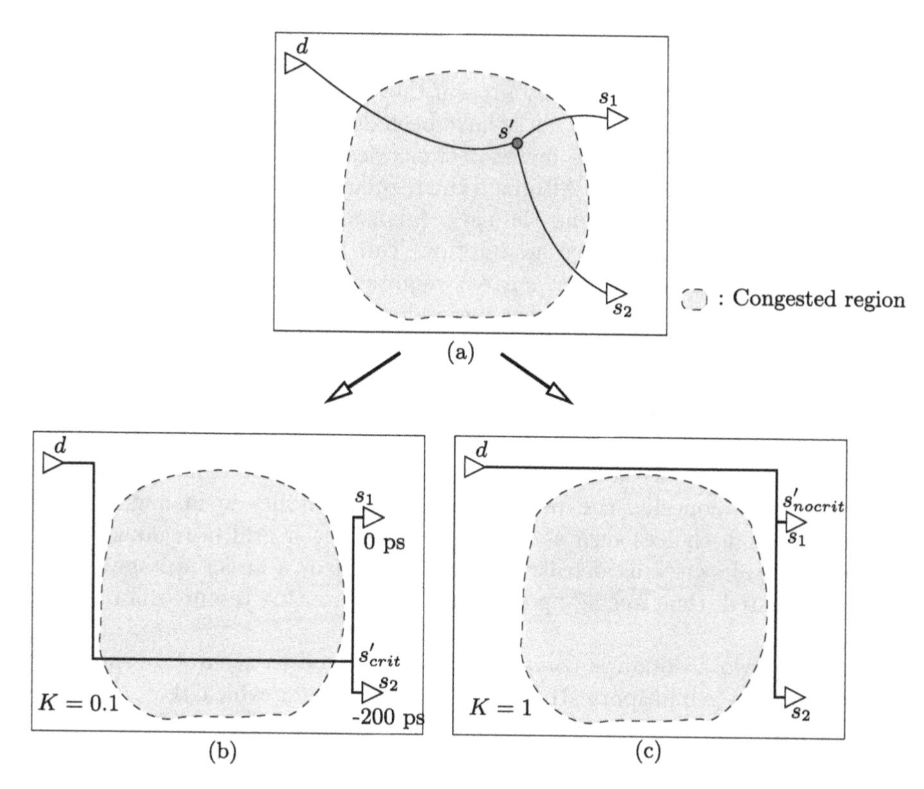

Fig. 4.17. An example illustrating the operation of BEN. (a) The input Steiner tree embedding. (b) A sample solution for a critical net. (The numbers at the sinks represent the required arrival times). (c) A sample solution for a non-critical net.

As an example, consider the net illustrated in Fig. 4.17. For the input Steiner tree embedding shown in Fig. 4.17(a), whose Steiner node s' lies within a congested region, the BEN algorithm can produce different embeddings and routings depending on the criticality of the net. Two sample solutions are shown for the cases when the net is critical and when the net is non-critical. Observe that the Steiner node s' can be moved from its original embedding in response to the cost function. Also, note that although the critical net is routed through the congested region because this routing improves the delay at the critical sink s_2, the non-critical net bypasses the congested region entirely.

4.4 Congestion Implications of Power Grid Design

One of the consequences of the increased device density and wire resistance caused due to process scaling is that power networks must now consume a significant fraction of the routing resources if they are to keep the voltage droop small. Traditionally, power networks have been designed as regular grids prior to signal routing (with coarse power network design being done even before early block implementation). Although the regularity of the grid simplifies the analysis of the power grid, and the early finalization of its design simplifies the design methodology, these assumptions can lead to some overdesign in the power grid, since the exact current requirements of a region cannot be accurately predicted until after that region has been fully implemented. This sometimes leads to more routing resource utilization by the power grid than may be absolutely necessary. Although this has not been a serious problem in the past, the increasing fraction of routing resources required by the power grid as well as the increasing congestion encountered during signal routing is leading to renewed research interest in the codesign of signal and power networks. Such codesign can provide increased flexibility in managing congestion, through choices such as a locally sparse power grid in regions where either the local switching density is not very high or a larger voltage droop can be tolerated, thus freeing up some additional routing resources for signal routing.

Power network design is complicated by the need to accurately simulate the network as a multiport RLC network - a task for which there are few good alternatives to SPICE-level simulations (although there has been much recent work in speeding up these simulations through the use of hierarchical or region-based schemes and power grid macromodels). Although simple delay abstractions such as lumped RC or Elmore delay models can be used within the inner loops of routers, there are no such simplifications for power network design. This makes the true codesign of signal and power networks a challenging problem. There are several existing approaches to the codesign problem, two of which are described in the remainder of this section. The first [SG03] of these gets around the problem of power grid analysis complexity by using a "guaranteed correct" power pitch abstraction while designing a non-uniform power grid without any on-the-fly analysis, whereas the second approach [SHS+02] adopts only a loose integration between power grid design and signal routing.

4.4.1 Integrated Power Network and Signal Shield Design

The work described in [SG03] is targeted towards high-end designs in which a significant number of signal nets require shielding or spacing. Spacing constraints on nets can be further translated into equivalent shielding constraints;

this translation results in no area overhead, improved noise characteristics and delay predictability, and only a small delay penalty (due to the additional non-switching capacitance of the shield). Although nets requiring shields have traditionally been routed opportunistically next to pre-existing power grid wires in order to avoid the need for a separate shield (thus saving a routing track), the work in [SG03] postpones the detailed implementation of the power grid within a block until after the shields required within that block have been laid out, and then tries to extract a viable non-uniform power grid from this shield network using an adaptive power routing algorithm. This approach thus creates additional flexibility in the design of the power grid, leading to a reduction in the total number of tracks required for the shield and power grid network. The effectiveness of this approach has been demonstrated on a leading edge microprocessor in high volume production.

In order to avoid expensive on-the-fly analysis of the power grid, this work relies on the abstraction of a *power pitch*, which is the maximum separation between successive power lines that guarantees acceptable voltage droop, inductive noise, and electromigration reliability. This pitch can be obtained for each block in a preprocessing phase, and depends on the local switching density of the block. The input to the adaptive power routing algorithm is a placed and partially routed block in which the major trunks of all the shielded signals have been routed, along with their shields, with the track assignment having been done so as to maximize the sharing of the shields among the shielded signals. However, at this stage, no power routing or non-critical signal routing has yet been done. Thus, the tracks which would have been used up for the local power grid in traditional methodologies have instead been made available for critical signal routing, in the hope that the shields required by these signals would take care of much of the power delivery requirements.

The adaptive power routing algorithm is described formally as Algorithm 6. Starting from one end of the design block, this algorithm first looks at the track one power pitch beyond the current track. Thus, in the example depicted in Fig. 4.18, if the algorithm is currently at track a, it first looks at track b that is P units beyond a (where P is the power pitch). If it is not able to find a shield there that can be reused for the power grid, it searches backwards until it finds a shield (at track b' in the example) or reaches the current track. If no shield is found, it then starts looking backward again for a vacant track starting from the track one power pitch beyond the current track and ending at the current track, and adds a power line to the first such vacant track that it finds. Once a shield or a vacant track has been found, the algorithm repeats the above procedure starting from the track P units beyond the selected track (*i.e.*, P units beyond track b' (and not b) in the example). The polarities of the shields that are reused for the power grid are set to alternate between V_{dd} and V_{ss} across the design block, whereas those for the remaining shields are set arbitrarily. In spite of its simplicity, this algorithm is provably optimal in the number of additional power lines that it adds in order to complete the power grid with a specified power pitch.

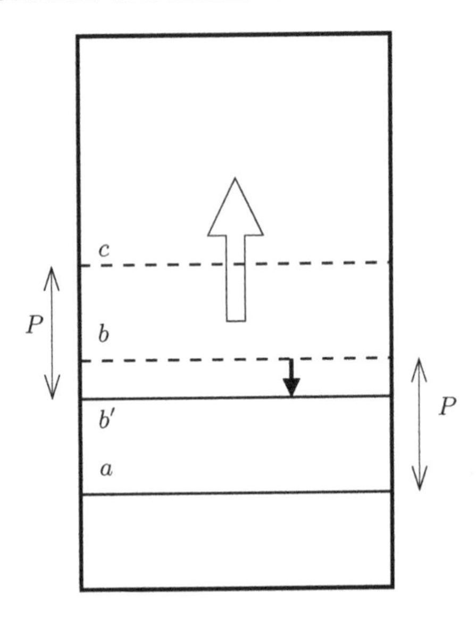

Fig. 4.18. Illustration of the operation of the adaptive power routing algorithm.

This adaptive power routing algorithm is applicable primarily to the detailed implementation of the power routing on the mid-level metal layers for design blocks that contain many signals that require shielding. The non-uniformity of the grid implies that the power lines within adjacent blocks may not match with each other. Since connecting them together through single layer doglegs can have a significant adverse impact on the local routability, these non-uniform local grids are left disconnected. This is usually not a problem since a large fraction of the current flow between these local mid-level layer grids for different blocks is through the upper metal layers because of their lower resistance, even if the local grids are connected together on the mid-level layers. Furthermore, unlike the upper metal layers where the power lines are quite wide, it is often preferable to keep the power grid on the mid-level layers quite fine-grained, with power lines whose width is equal to or slightly larger than the minimum permissible width for signal lines on that layer. Thus, the shields can be wide enough to be reused as power lines without significant area penalty. On the other hand, this approach does not yield large benefits in designs that are not aggressive and do not require much shielding or spacing of signal nets.

4.4.2 Signal and Power Network Codesign

The work described in [SHS+02] attempts a more comprehensive integration of power and signal routing than [SG03], and is therefore applicable to a wider class of designs. It starts with a dense power grid that is guaranteed to

Algorithm 6 Adaptive power routing to extract a local power grid with power pitch P from a given shield network

1: $target \leftarrow P$
2: **while** $target < maxTrackIndex$ **do**
3: // Look for next shield or vacant track
4: $track \leftarrow target$
5: **while** no shield found and $track > target - P$ **do**
6: **if** $track$ contains a shield **then**
7: Reuse shield in $track$ for power grid
8: **else**
9: $track \leftarrow track - 1$
10: **end if**
11: **end while**
12: **if** no shield found **then**
13: $track \leftarrow target$
14: **while** no vacant track found and $track > target - P$ **do**
15: **if** $track$ is vacant **then**
16: Add a power line to $track$
17: **else**
18: $track \leftarrow track - 1$
19: **end if**
20: **end while**
21: **end if**
22: **if** no shield or vacant track found **then**
23: Report local power grid failure around $target$ and exit
24: **end if**
25: $target \leftarrow track + P$
26: **end while**

meet all the constraints on the power delivery network, and then iteratively sparsifies it in high congestion regions, sizing the remaining power grid wires to compensate for the deleted wires and meet the power delivery constraints, generating a power grid similar to the example depicted in Fig. 4.19.

The algorithm begins with the generation of a congestion map under the assumption of a dense and uniform power grid. This map can be obtained using any of the various congestion estimation techniques discussed in Chapter 2 (such as probabilistic congestion maps or fast global routers); the implementation in [SHS+02] uses the framework of [AHS+03], briefly discussed in Section 4.3.2. More specifically, it first performs a coarse global routing for all the signal nets in a congestion-oblivious fashion, generating Steiner topologies for multipin nets using the hybrid Prim-Dijkstra algorithm from [AHH+95]. This is followed by an iterative congestion-driven rip-up and reroute stage during which the cost of routing a wire through a boundary edge of a global routing bin is inversely proportional to the dynamically updated routing capacity remaining on that edge, and the regeneration of congestion-aware tree topologies is done using the Steiner min-max tree algorithm from [CS90] (that

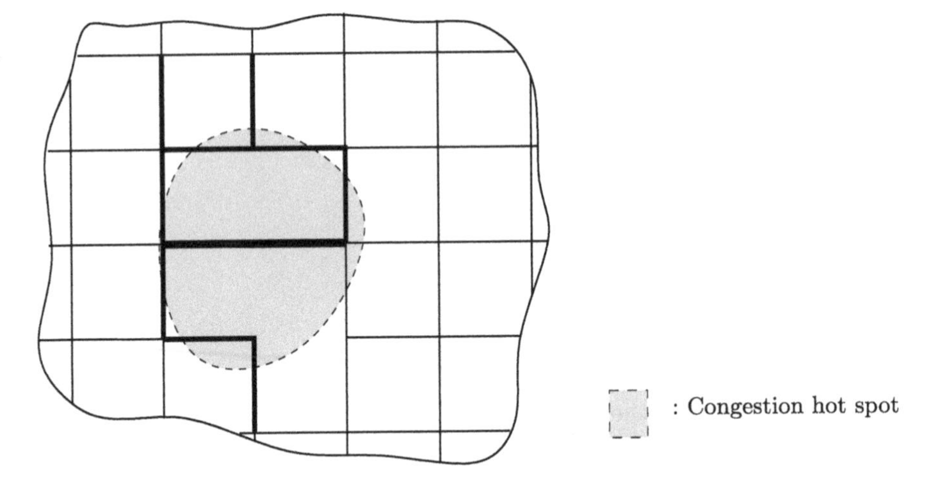

Fig. 4.19. Deletion of some power grid wire segments and sizing of others in response to routing congestion.

was discussed in Section 4.1.1). This signal routing stage is used to identify the high congestion regions, which are then used to direct the local sparsification of the power grid in those regions.

However, not all power grid wires in congested regions are candidates for deletion. Each segment of a power grid wire that lies within a single bin is considered separately. If the worst-case voltage droop on some wire segment is greater than some pre-specified threshold, it is marked as critical and is not considered for deletion. For each of the remaining power wire segments, a criticality metric is defined as the reciprocal of the root-mean-squared (RMS) distance of that segment from power grid nodes with voltage droop greater than some threshold, lying within a given neighborhood of the bin containing that segment. Thus, if c_b represents the center of bin b, the criticality $Crit_p$ of a vertical wire p crossing a horizontal bin boundary at x_p is defined as:

$$Crit_p = \frac{K_{p,\Delta}}{\sqrt{\sum_{i:critical,|x_p - c_i| < \Delta} (x_p - x_i)^2}},$$

where $K_{p,\Delta}$ is the total number of noisy nodes lying within some Δ-neighborhood of the bin containing p. The criticality of horizontal power grid segments is computed in an analogous manner. (The implementation of [SHS+02] uses a Δ of 1.5–2 bin lengths). Thus, wire segments that are close to voltage droop hot spots end up with a larger criticality metric.

Next, the bin edges are sorted in terms of their routing overflow, and, for each bin edge, the non-critical power grid wire segments crossing that edge are sorted by their criticality metric. Deletion of wire segments is done by selecting the least critical wire segments in the most congested bin edges first,

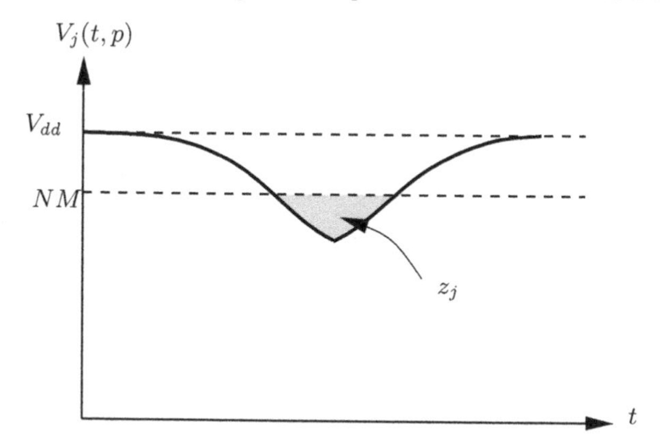

Fig. 4.20. Computation of the power grid noise metric from [CHV98] at a node.

incrementally updating the routing overflows (and the sorted list of bin edges) after each deletion. After a certain number of power grid wire segments have been deleted, the remainder of the wires in the power grid are widened in order to compensate for the deleted wires, and their criticalities are recomputed. This sizing of the power grid is done by modeling it as a constrained nonlinear program (NLP) as follows:

$$\min Area = \sum_{j=1}^{N_{wire}} l_j \times w_j,$$

subject to:

$$Z(w_1, \ldots, w_{N_{wire}}) < \epsilon,$$

$$w_{\min} \leq w_j \leq w_{\max},$$

for each $j = 1, \ldots, N_{wire}$, where N_{wire} is the number of wire segments in the power grid, w_j and l_j are, respectively, the width and length of the jth wire segment, Z is the voltage droop metric from [CHV98] defined as the sum of the individual node power grid noise values (z_j's) computed by integrating the power grid noise violation at each node (represented by the shaded area in Fig 4.20), and ϵ is a small constant that bounds this metric. The metric Z can be obtained by a transient analysis of the power grid circuit, and its sensitivities with respect to the widths of the different power grid wire segments can be obtained using the adjoint method as discussed in [CHV98]. The nonlinear program for power grid sizing can be solved using any standard sequential quadratic programming solver; an approximate solution that can be obtained quickly usually suffices. The overall flow of this signal and power network codesign scheme is summarized in Algorithm 7.

Algorithm 7 Congestion-driven codesign of the signal and power networks

1: **while** previous iteration improved congestion, but significant congestion overflow still exists **do**
2: Generate congestion map
3: Perform transient simulation of power grid, calculating sensitivities to wire segment widths
4: **while** number of deleted power grid wire segments $< \alpha N_{wire}$ **do**
5: $e \leftarrow$ routing bin edge with maximum overflow
6: Delete the least critical wire segment crossing e
7: Update congestion overflow in e
8: **end while**
9: Set up and solve NLP to size remaining wire segments in power grid
10: **end while**

4.5 Congestion-aware Interconnect Noise Management

With wires becoming increasingly resistive at each successive process technology node, process engineers have attempted to ameliorate the problem by increasing the aspect ratio of the wires, since the resulting tall wires do not degrade routability (because they do not require a wider pitch). However, as was discussed in Chapter 1, tall wires aggravate the problem of interconnect crosstalk. This manifests itself not only through functional failures causing the inversion of data bits stored in sequential elements or produced at the primary outputs, but also in widened switching windows for the signals resulting in a harder design convergence problem.

Although the optimization of interconnect noise has been studied extensively, most of the early works in this area did not explicitly consider the congestion impact of the various proposed noise fixes. Although some of these fixes (such as net ordering and gate sizing) do not have a significant impact on the routing congestion, the feasibility of many other important fixes such as the shielding or spacing of the wires, buffer insertion, or the rerouting of the problem nets has a strong dependence on the local congestion. Yet, just like the early works on interconnect buffering, many of the early publications on interconnect noise optimization also ignored the interdependence of this problem with routing congestion, and focused on studying the problem in isolation at the level of individual nets.

Most modern commercial global routers support rudimentary shield insertion and spacing, as well as more explicit crosstalk control by prohibiting the extended routing of pairs of mutually sensitive nets in adjacent tracks. The integrated power network and signal shield design algorithm [SG03] discussed in Section 4.4.1 attempted a congestion-aware allocation of shields in two ways, namely, (i) by maximizing the reuse of the local power grid for shielding by making it non-uniform, and (ii) by maximizing the sharing of shields during net ordering and track assignment. More recently, there have been a few approaches [XH05] [ZS04] that have incorporated shield allocation and other noise fixes into full-fledged global routers.

4.5.1 Congestion-aware Shield Synthesis for RLC Noise

The work in [XH05] attempts to optimize the routing not merely for purely capacitive crosstalk but also for the inductive noise between the nets. It relies on the empirically fitted K_{eff} model for inductive noise that allows the total inductive noise $K_{i,t}$ experienced by a net N_i in a particular routing region R_t to be expressed as the weighted sum of its coupling coefficients with all the other locally routed nets that it is sensitive to. This noise model is further extended to a *length scaled* K_{eff} model that allows the total inductive noise experienced by a net N_i to be expressed as $\sum_t K_{i,t}l_t$, where the summation is over all t such that N_i is routed through the region R_t, and l_t is the length of the routing of N_i in R_t. The routing regions are defined in terms of the power grid, so that each routing region guarantees local return paths for the induced currents.The linear composition of the local noise in each segment of a net to obtain its total noise using the K_{eff} model allows for the budgeting of the total noise slack at each sink of a net into local noise slacks for the net within each of the regions that it is routed through.

Within each region, the number of shields required by the local nets is estimated using an empirical closed form linear expression that models the shielding requirements of the simulated annealing based SINO (Shield Insertion and Net Ordering) algorithm for the optimization of RLC noise within a routing region. The shield count estimation within a region allows for the computation of the expected routing overflow in that region after taking the subsequent shield synthesis into account, without having to actually synthesize the shields. This procedure is embedded into a global router; the implementation in [XH05] uses a global router based on iterative deletion (discussed in Section 4.1.6), although any other global routing scheme can be used as well. The global router can then carry out the routing of all the nets, taking into account the routing demand not only from the signal nets but also their shields (that are yet to be synthesized), dynamically updating the shield count estimation (and therefore the routing congestion) in each region as the global routing proceeds. Finally, once all the nets have been global routed, their shields are synthesized and track assignment is carried out within each routing region using SINO.

Although this work is interesting because it relies on a fast shielding overhead estimator to compute local congestion during global routing, it suffers from a somewhat simplistic model for the coupling between the nets. Furthermore, interconnect noise in today's designs is usually dominated by capacitive coupling, although inductive coupling may grow in importance for wide wires at the higher frequencies of future designs.

4.5.2 Integrated Congestion-aware Shielding and Buffering

The more traditional capacitive interconnect crosstalk problem is tackled within a global routing framework in [ZS04] through integrated shielding and buffering. This work relies on the "Elmore noise model" proposed for coupled interconnects in [Dev97]. Given a wire segment (i, j) of length l_{ij}, with i being the upstream node, the noise current $I_n(i)$ and noise margin $NM(i)$ at the node i can be computed using this noise metric as follows:

$$I_n(i) = I_n(j) + l_{ij} c_c s_a,$$
$$NM(i) = NM(j) - r l_{ij} (\frac{1}{2} c_c l_{ij} s_a + I_n(j)),$$

where c_c and r are the coupling capacitance and resistance, respectively, of a unit length of the wire, and s_a is the rate at which the signal in the aggressor net coupled to the wire segment (i, j) is switching. These equations allow for the bottom-up computation of the coupled noise in a net starting from its sinks. The net is considered to have a noise failure if the total induced current I_n at its driver is greater than NM/R_d, where NM is the maximum acceptable noise voltage threshold at the driver, and R_d is the effective driver resistance.

Observe that the functional form of the above equations is very similar to that of the Elmore delay model [RPH83] (with noise current in place of downstream capacitance, noise margin in place of delay slack, and an extra term for aggressor signal slew). This suggests the use of a procedure similar in spirit to the well-known dynamic programming based van Ginneken algorithm [Van90] for buffer insertion in a given net under the Elmore delay model, in order to instead optimize for interconnect noise in a given net. Such an algorithm for noise-aware interconnect buffering was presented in [ADQ99].

The algorithm proposed in [ZS04] extends that work to incorporate shielding as well as routability concerns. It achieves this by traversing each problem net bottom up, choosing whether or not to insert a buffer and whether to provide single-sided, double-sided or no shielding at each node. At each node, these choices result in six possible configurations at the parent edge of the node, which can then be combined with the previously generated partial solutions for the noise-optimized routing tree rooted at the node. A partial solution at a node that is inferior to any other partial solution at that node is pruned away, and the remaining solutions are propagated upstream. The algorithm also enforces a maximum length constraint on the total interconnect length that can be driven by a buffer, by bucketing the potential solutions at any node by the distance to the closest downstream buffers (in a manner similar to [AHS+03]). The cost of a partial solution captures the buffer congestion cost of each global bin along the partial routing tree, in addition to the routing congestion cost of any shields that may have been added.

This procedure is preceded by congestion-driven global routing, implemented in [ZS04] using a framework similar to that proposed in [AHS+03] (which was briefly discussed in Section 4.3.2). This global routing is followed by an iterative loop in which the simultaneous buffering and shielding procedure described above is followed by a local rip-up and reroute of nets that still have noise violations, going through all the nets in a fixed order. The iterations are continued until all noise and congestion problems are resolved, or until there is no further improvement.

Although the noise model of [Dev97] is known to be pessimistic, the work in [ZS04] demonstrates that this model exhibits good fidelity with SPICE simulations. Furthermore, the pessimism of the noise model is countered by inflating the noise margin thresholds empirically at the gates.

4.6 Final Remarks

Since the primary goal of routing has traditionally been route completion, there has been considerable work over the years on improving the route completion rate of routing algorithms. Global routing focuses on spreading the congestion uniformly across the layout, so that routing overflows are minimized, whereas detailed routing cleans up any remaining routing overflows and addresses pin accessibility issues, by looking intensively at a small piece of the layout. In this chapter, we have seen that the biggest challenge to effective global routing is the net ordering problem, and that the standard approach to minimize its effect is through sophisticated rip-up and reroute heuristics, often augmented with hierarchical schemes. This problem has also spurred considerable research in more "concurrent" global routing techniques such as those using multicommodity flows, but these approaches do not yet scale up to today's large problem sizes as well as the more standard methods.

With process scaling, the resistance of wires tends to increase, resulting in the increasing importance of several electrical effects that have an indirect effect on routing congestion. Thus, power grids require increased routing resources in order to control the voltage droop, whereas signal nets require shielding and buffering in order to avoid excessive wire delays, poor signal slews, noise failures and large switching windows. As a consequence, the routing congestion problem becomes even more severe. The latest generation of research in tackling these electrical problems does so within the context of the physical environment, making the desired fixes locally without ignoring their impact on the routability of the design. As more and more electrical and manufacturing effects start becoming prominent, they too will need to be addressed in a similar holistic fashion.

References

[Alb00] Albrecht, C., Provably good global routing by a new approximation algorithm for multicommodity flow, *Proceedings of the International Symposium on Physical Design*, pp. 19–25, 2000.

[AD97] Alpert, C. J., and Devgan, A., Wire segmenting for improved buffer insertion, *Proceedings of the Design Automation Conference*, pp. 588–593, 1997.

[ADQ99] Alpert, C. J., Devgan, A., and Quay, S. T., Buffer insertion for noise and delay optimization, *IEEE Transactions on Computer-Aided Design of Integrated Circuits and Systems* 18(11), pp. 1633–1645, Nov. 1999.

[AGH+02] Alpert, C. J., Gandham, G., Hrkic, M., Hu, J., Kahng, A. B., Lillis, J., Liu, B., Quay, S. T., Sapatnekar, S. S., and Sullivan, A. J., Buffered Steiner trees for difficult instances, *IEEE Transactions on Computer-Aided Design of Integrated Circuits and Systems* 21(1), pp. 3–14, Jan. 2002.

[AGH+04] Alpert, C. J., Gandham, G., Hrkic, M., Hu, J., Quay, S. T., and Sze, C. N., Porosity-aware buffered Steiner tree construction, *IEEE Transactions on Computer-Aided Design of Integrated Circuits and Systems* 23(4), pp. 517–526, April 2004.

[AHH+04] Alpert, C. J., Hrkic, M., Hu, J., and Quay, S. T., Fast and flexible buffer trees that navigate the physical layout environment, *Proceedings of the Design Automation Conference*, pp. 24–29, 2004.

[AHH+95] Alpert, C. J., Hu, T. C., Huang, J. C., Kahng, A. B., and Karger, D., Prim-Dijkstra tradeoffs for improved performance-driven routing tree design, *IEEE Transactions on Computer-Aided Design of Integrated Circuits and Systems* 14(7), pp. 890–896, July 1995.

[AHS+03] Alpert, C. J., Hu, J., Sapatnekar, S. S., and Villarrubia, P. G., A practical methodology for early buffer and wire resource allocation, *IEEE Transactions on Computer-Aided Design of Integrated Circuits and Systems* 22(5), pp. 573–583, May 2003.

[BP83] Burstein, M., and Pelavin, R., Hierarchical wire routing, *IEEE Transactions on Computer-Aided Design of Integrated Circuits and Systems* 2(4), pp. 223–234, Oct. 1983.

[CLC96] Carden, R. C., Li, J., and Cheng, C.-K., A global router with a theoretical bound on the optimal solution, *IEEE Transactions on Computer-Aided Design of Integrated Circuits and Systems* 15(2), pp. 208–216, Feb. 1996.

[CL04] Chang, Y.-W., and Lin, S.-P., MR: A new framework for multilevel full-chip routing, *IEEE Transactions on Computer-Aided Design of Integrated Circuits and Systems* 23(5), pp. 793–800, May 2004.

[CS90] Chiang, C., and Sarrafzadeh, M., Global routing based on Steiner min-max trees, *IEEE Transactions on Computer-Aided Design of Integrated Circuits and Systems* 9(12), pp. 1318–1325, Dec. 1990.

[CWS94] Chiang, C., Wong, C. K., and Sarrafzadeh, M., A weighted Steiner tree based global router with simultaneous length and density minimization, *IEEE Transactions on Computer-Aided Design of Integrated Circuits and Systems* 13(12), pp. 1461–1469, Dec. 1994.

[CS98] Cho, J. D., and Sarrafzadeh, M., Four-bend top-down global routing, *IEEE Transactions on Computer-Aided Design of Integrated Circuits and Systems* 17(9), pp. 793–802, Sep. 1998.

[Con97] Cong, J., Challenges and opportunities for design innovations in nanometer technologies, *Frontiers of Semiconductor Research: A Collection of SRC Working Papers*, available at `http://www.src.org/prg_mgmt/frontier.dgw`, 1997.

[CFX+05] Cong, J., Fang, J., Xie, M., and Zhang, Y., MARS – A multilevel full-chip gridless routing system, *IEEE Transactions on Computer-Aided Design of Integrated Circuits and Systems* 24(3), pp. 382–394, Mar. 2005.

[CKP99] Cong, J., Kong, T., and Pan, D. Z., Buffer block planning for interconnect driven floorplanning, *Proceedings of the International Conference on Computer-Aided Design*, pp. 358–363, 1999.

[CP92] Cong, J., and Preas, B., A new algorithm for standard cell global routing, *Integration — The VLSI Journal* 14(1), pp. 49–65, 1992.

[CHV98] Conn, A. R., Haring, R. A., and Visweswariah, C., Noise considerations in circuit optimizations, *Proceedings of the International Conference on Computer-Aided Design*, pp. 220–227, 1998.

[Dev97] Devgan, A., Efficient coupled noise estimation for on-chip interconnect, *Proceedings of the International Conference on Computer-Aided Design*, pp. 147–151, 1997.

[Elm48] Elmore, W. C., The transient response of damped linear network with particular regard to wideband amplifiers, *Journal of Applied Physics* 19(1), pp. 55–63, Jan. 1948.

[GK98] Garg, N., and Könemann, J., Faster and simpler algorithms for multicommodity flow and other fractional packing problems, *Proceedings of the Symposium on Foundations of Computer Science*, pp. 300–309, 1998.

[HT95] Hayashi, M., and Tsukiyama, S., A hybrid hierarchical approach for multilayer global routing, *Proceedings of the European Design and Test Conference*, pp. 492–496, 1995.

[HL91] Heisterman, J., and Lengauer, T., The efficient solution of integer programs for hierarchical global routing, *IEEE Transactions on Computer-Aided Design of Integrated Circuits and Systems* 10(6), pp. 748–753, June 1991.

[Hig69] Hightower, D. W., A solution to line routing problems on the continuous plane, *Proceedings of the Design Automation Workshop*, pp. 1–24, 1969.

[HS00] Hu, J., and Sapatnekar, S. S., A timing-constrained algorithm for simultaneous global routing of multiple nets, *Proceedings of the International Conference on Computer-Aided Design*, pp. 99–103, 2000.

[HS01] Hu, J., and Sapatnekar, S. S., A survey on multi-net global routing for integrated circuits, *Integration — The VLSI Journal* 31(1), pp. 1–49, Nov. 2001.

[KGV83] Kirkpatrick, S., Gelatt, C. D., and Vecchi, M. P., Optimization by simulated annealing, *Science* 220, pp.671–680, 1983.

[Lee61] Lee, C. Y., An algorithm for path connection and its applications, *IRE Transactions on Electronic Computers* EC-10(3), pp. 346–365, 1961.

[LS91] Lee, K. Y., and Sechen, C., A global router for sea-of-gates circuits, *Proceedings of the European Design Automation Conference*, pp. 242–247, 1991.

[LCL96] Lillis, J., Cheng, C.-K., and Lin, T.-T. Y., Optimal wire sizing and buffer insertion for low power and a generalized delay model, *IEEE Journal on Solid-state Circuits* 31(3), pp. 437–447, 1996.

[LHT90] Lin, Y.-L., Hsu, Y.-C., and Tsai, F.-S., Hybrid routing, *IEEE Transactions on Computer-Aided Design of Integrated Circuits and Systems* 9(2), pp. 151–157, Feb. 1990.

[Mar84] Marek-Sadowska, M., Global router for gate array, *Proceedings of the International Conference on Computer Design*, pp. 332–337, 1984.

[ML90] Meixner, G., and Lauther, U., A new global router based on a flow model and linear assignment, *Proceedings of the International Conference on Computer-Aided Design*, pp. 44–47, 1990.

[Nai87] Nair, R., A simple yet effective technique for global wiring, *IEEE Transactions on Computer-Aided Design of Integrated Circuits and Systems* 6(2), pp. 165–172, Oct. 1987.

[NHL+82] Nair, R., Hong, S. J., Liles, S., and Villani, R., Global wiring on a wire routing machine, *Proceedings of the Design Automation Conference*, pp. 224–231, 1982.

[RPH83] Rubenstein, J., Penfield, P., and Horowitz, M. A., Signal delay in RC tree networks, *IEEE Transactions on Computer-Aided Design of Integrated Circuits and Systems* CAD-2, pp. 202–211, July 1983.

[SK01] Sarkar, P., and Koh, C.-K., Routability-driven repeater block planning for interconnect-centric floorplanning, *IEEE Transactions on Computer-Aided Design of Integrated Circuits and Systems* 20(5), pp. 660–671, May 2001.

[Sax06] Saxena, P., The scaling of interconnect buffer needs, *Proceedings of the International Workshop on System Level Interconnect Prediction*, pp. 109–112, 2006.

[SG03] Saxena, P., and Gupta, S., On integrating power and signal routing for shield count minimization in congested regions, *IEEE Transactions on Computer-Aided Design of Integrated Circuits and Systems* 22(4), pp. 437–445, April 2003.

[SL01] Saxena, P., and Liu, C. L., Optimization of the maximum delay of global interconnects during layer assignment, *IEEE Transactions on Computer-Aided Design of Integrated Circuits and Systems* 20(4), pp. 503–515, April 2001.

[SMC+04] Saxena, P., Menezes, N., Cocchini, P., and Kirkpatrick, D. A., Repeater scaling and its impact on CAD, *IEEE Transactions on Computer-Aided Design of Integrated Circuits and Systems* 23(4), pp. 451–463, April 2004.

[SS86] Sechen, C., and Sangiovanni-Vincentelli, A., TimberWolf 3.2: A new standard cell placement and global routing package, *Proceedings of the Design Automation Conference*, pp. 432–439, 1986.

[SM90] Shahrokhi, F., and Matula, D. W., The maximum concurrent flow problem, *Journal of the ACM* 37(2), pp. 318–334, 1990.

[SK87] Shragowitz, E., and Keel, S., A global router based on a multicommodity flow model, *Integration — The VLSI Journal* 5(1), pp. 3–16, March 1987.

[SHS+02] Su, H., Hu, J., Sapatnekar, S. S., and Nassif, S. R., Congestion-driven codesign of power and signal networks, *Proceedings of the Design Automation Conference*, pp. 64–69, 2002.

[SHA04] Sze, C. N., Hu, J., and Alpert, C. J., A place and route aware buffered Steiner tree construction, *Proceedings of the Asia and South Pacific Design Automation Conference*, pp. 355–360, 2004.

[TT83] Ting, B. S., and Tien, B. N., Routing techniques for gate array, *IEEE Transactions on Computer-Aided Design of Integrated Circuits and Systems* 2(4), pp. 301–312, Oct. 1983.

[Van90] van Ginneken, L. P. P. P., Buffer placement in distributed RC-tree networks for minimal Elmore delay, *Proceedings of the International Symposium on Circuits and Systems*, pp. 865–868, 1990.

[VK83] Vecchi, M. P., and Kirkpatrick, S., Global wiring by simulated annealing, *IEEE Transactions on Computer-Aided Design of Integrated Circuits and Systems* 2(4), pp. 215–222, Oct. 1983.

[XH05] Xiong, J., and He, L., Extended global routing with RLC crosstalk constraints, *IEEE Transactions on Very Large Scale Integration Systems* 13(3), pp. 319–329, Mar. 2005.

[ZS04] Zhang, T., and Sapatnekar, S., Simultaneous shield and buffer insertion for crosstalk noise reduction in global routing, *Proceedings of the International Conference on Computer Design*, pp. 93–98, 2004.

5 CONGESTION OPTIMIZATION DURING PLACEMENT

Building congestion awareness during placement is a very effective way of improving the routability of a design, since the routing solution space available during placement is considerably larger than the one that can be explored during routing. Indeed, the optimization of routing congestion during placement is what usually distinguishes congestion-aware versions of modern physical synthesis flows from their congestion-oblivious versions. However, the metric that has traditionally been optimized during placement is the sum of the (estimated) wirelengths of all the nets in the design. Even performance-driven placement algorithms usually target merely a weighted sum of the wirelengths of the nets. While this wirelength minimization also reduces the average routing congestion to some extent, congestion is inherently a local problem. The thresholded nature of routing overflow (*i.e.*, the lack of contribution of wires passing through an uncongested global routing cell to the overall congestion) often means that a minimal estimated wirelength design may not only have a very high maximum congestion, but may also have a higher average congestion than some other placement with a longer total estimated wirelength. Indeed, it is quite possible that a placement with minimal estimated wirelength may be unroutable, unlike some other congestion-aware placement with a longer estimated wirelength.

As a toy illustration of the problem described above, consider Fig. 5.1. This figure depicts a block containing two cells c_1 and c_2, each of which is to be connected to two pins on the boundary of the block. The block also includes a U-shaped blockage that divides the placement area into two regions, R_1 and R_2, that lie above and below the blockage, respectively. Let the channels between the blockage and the boundaries of the block have a routing capacity of one track each. Then, the placement which minimizes the total wirelength (as estimated using the half-rectangle perimeters (HRPM) of bounding boxes of the nets) will place both the cells c_1 and c_2 within the region R_1 and will be unroutable; however, moving either of the cells to the region R_2 will permit routing completion, albeit at an increased wirelength, as shown in Fig. 5.1(b).

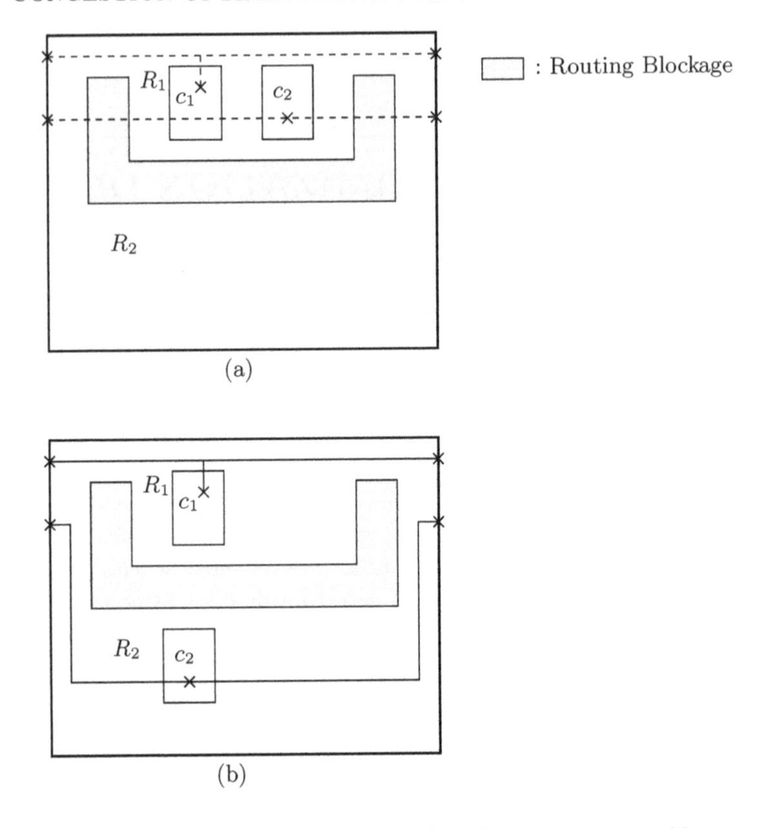

Fig. 5.1. A placement that minimizes wirelength may be unroutable.

Congestion-aware placement has been a fertile field of research over the last decade, and numerous such techniques have been proposed in the literature and implemented in commercial physical synthesis tools. However, based on when they are applicable during the layout flow, most of these techniques can be broadly classified into three categories, namely,

1. optimizing the routing congestion as a post-processing step after the global placement (a step that is often integrated into the process of detailed placement and legalization),
2. interleaving between global placement operations and routing congestion estimation and optimization, and,
3. incorporating routing congestion optimization directly into the global placement algorithm.

These three categories are discussed in depth in Sections 5.2, 5.3, and 5.4, respectively.

As is usual with the optimization of any layout metric at different stages in the physical synthesis flow, techniques applied further upstream can explore a richer optimization space but have to work with poorer estimates of the metric. Thus, congestion estimates during early placement are inaccurate not only because the locations of the cells have not yet been fixed, but also because they do not capture the effect of subsequent netlist transformations such as gate resizing, buffer insertion and local resynthesis, or layout transformations such as clock tree layout, shield insertion, and wire sizing. On the other hand, although the post-processing methods can rely on more accurate congestion estimates (especially at fine granularities) than those integrated into the global placement, they suffer from a somewhat decreased flexibility in the extent of permissible design change and optimization headroom.

The other major concern in making placement congestion-aware is the runtime overhead for estimating and optimizing congestion. Ideally, every global placement move should be followed by a complete analysis of its impact on the routing congestion map obtained by actually running the router on the current placement of the design. However, this is unaffordable even with the use of incremental routing. Therefore, different techniques try to reduce the runtime overhead either through fast (albeit less accurate) congestion estimation models (such as those discussed in Chapter 2), or through avoiding the analysis of the congestion impact of individual placement moves by postponing this analysis until a certain number of moves has been carried out or some predetermined placement convergence criteria have been met. As a consequence, the actual congestion gains obtained due to some move carried out during early global placement may be much smaller than anticipated.

Before we dive into the details of various placement-level techniques for congestion mitigation (in Sections 5.2, 5.3 and 5.4), it is useful to digress briefly for a quick overview of placement techniques currently in vogue.

5.1 A Placement Primer

While a detailed discussion of the theory and implementation of the various placement methods proposed in the literature and applied in practice is outside the scope of this chapter, a brief synopsis of these techniques can help set the context for our discussion of their modification for congestion awareness. A more extensive survey of the current state of the art in placement can be found in [CSX+05].

The problem of placing the cells in a netlist inside a specified region in a way that the cells do not overlap with each other and some cost function (usually, the total estimated wirelength of the design) is minimized has been studied in depth for several decades. The established paradigm divides this task into two stages: (i) *global placement* that tries to exploit the overall structure of the netlist to obtain a placement that minimizes the cost function, even if the cells are not overlap-free, and (ii) *legalization* (also known as *detailed placement*) that uses local moves to eliminate the overlaps between the cells in the global-placed design and arrange them into the underlying layout architecture (such as fixed-height rows) while minimizing the deterioration in the cost function.

While many optimization techniques have been proposed for placement, most of these techniques[1] can be classified into:

- Analytical methods,
- Top-down partitioning-based methods,
- Multilevel methods, and,
- Simulated annealing and other move-based methods.

Each of these classes of placement techniques have their own strengths and weaknesses, that are discussed next. Most good placers combine techniques from multiple classes in an attempt to achieve high quality, robust placements within a reasonable computation time. Furthermore, most of today's industrial placers operate primarily in a timing-driven mode which allows the nets lying on timing-critical paths to be weighted more heavily than other nets.

5.1.1 Analytical Placement

In its simplest version, analytical placement models the netlist as a system of springs or resistors. First introduced in PROUD [CK84,THK88], this technique was refined in GORDIAN [KSJ+91]. The traditional objective function Φ used in such approaches seeks to minimize the weighted sum of squared Euclidean distances of connected cells and can be expressed in matrix notation as:

$$\Phi = \frac{1}{2}\mathbf{p}^T C\mathbf{p} + \mathbf{d}^T\mathbf{p} + k, \tag{5.1}$$

where C is a $2n \times 2n$ symmetric positive definite matrix (referred to as the *Laplacian* of the netlist) that captures the connectivity of the netlist, $\mathbf{d}$ is a $2n$-dimensional vector representing the fixed pin connections, $\mathbf{p} = (x_1, \ldots, x_n, y_1, \ldots, y_n)^T$ is a $2n$-dimensional vector representing the placement of the circuit (n being the number of cells in the circuit, with cell c_i located at $\mathbf{p_i} = (x_i, y_i)$), and k is a constant. Φ can be minimized by setting:

$$\triangledown\Phi(\mathbf{p}) = \mathbf{0},$$

where $\triangledown$ is the gradient operator, resulting in the linear system:

$$C\mathbf{p} + \mathbf{d} = \mathbf{0}. \tag{5.2}$$

This formulation is equivalent to modeling the nets as springs using Hooke's Law and then calculating the state of equilibrium. Consequently, it is also referred to as *force-directed* placement.

[1] Other approaches to placement, such as those using linear programming, have also been proposed, but have not been very successful in practice.

While quadratic formulations for placement had been proposed earlier, GORDIAN's contribution was in using partitioning to spread cells rather than to reduce the size of the mathematical program, thus reducing the penalty due to poor choices during the early partitioning stages. This was achieved by adding center of gravity constraints for the cells in each of the partitions, namely,

$$\frac{1}{n_j} \sum_{c_i \in P_j} \mathbf{p_i} = \mathbf{p_j^{(0)}}, \tag{5.3}$$

where n_j is the number of cells in partition P_j, and $\mathbf{p_j^{(0)}}$ is the current center of gravity for the cells in this partition. Embedding these constraints into the unconstrained system reduces the number of cells that can be placed independently. The unconstrained system on the remaining independent variables (say, $\mathbf{p_I}$) can be expressed in vector notation as:

$$\Phi_I = \frac{1}{2}\mathbf{p_I}^T Z^T C Z \mathbf{p_I} + \mathbf{d_I}^T \mathbf{p_I} + k, \tag{5.4}$$

using the matrix Z, and minimized by solving:

$$Z^T C Z \mathbf{p_I} + \mathbf{d_I} = \mathbf{0}, \tag{5.5}$$

where,

$$\mathbf{d_I}^T = (C\mathbf{p}^{(0)} + \mathbf{d})^T Z,$$

and $\mathbf{p}^{(0)}$ is an assignment of the dependent location variables that satisfies the center of gravity constraints specified in Equation (5.3). Since all the equations in this formulation are separable along the x and y dimensions, GORDIAN alternates between adjusting the x-coordinates and the y-coordinates of the placement in successive iterations.

This analytical approach was extended in GORDIANL [SDJ91] which improved the wirelength further by optimizing for the (non-differentiable) "linear" sum-of-wirelengths objective function using nested iterations on the net weights within the traditional quadratic framework. The next major contribution was that of KRAFTWERK [EJ98], in which the spring analogy was extended further by iteratively creating a spreading force field, consisting of additional springs required to effectively spread the cells to cover the placement area. This formulation relied on certain mathematical conditions to ensure that the forces were easily computable, by recasting the formulation in the form of Poisson's equation. Recently, the quadratic objective function has been replaced by the *log-sum-exp* model for wirelength, yielding significant improvements in placement quality [NDS01,KW04]. On another front, FASTPLACE [CV04] has shown large runtime improvements by integrating several simple local and global heuristics for cell moves into the iterative quadratic programming framework for analytical placement.

Analytical placement techniques tend to be "stable". In other words, small changes in the input netlist usually do not result in dramatic changes in the resulting placement. This makes them an attractive option for real-world applications, where design convergence involves numerous small netlist changes to an almost-converged netlist and layout.

5.1.2 Top-down Partitioning-based Placement

Placement using recursive top-down partitioning was first explored in [Bre77] and [DK85]. Early works in this class of techniques relied on simple recursive bisection with a cut size objective function. The cut size improvement at any level of the recursion was carried out using variants of Fiduccia-Mattheyses [FM82] iterations for moving nodes from one partition to the other. Recent advances in fast hypergraph partitioning have helped create a new generation of high quality top-down partitioning-based placers such as CAPO [CKM00] (discussed further in Section 5.4.2) and FENGSHUI [YM01].

Given any region to be partitioned and a set of cells that have already been assigned to that region, these techniques first select a horizontal or vertical cut line for that region. Then, subsets of cells are moved across that cut line in a way that reduces the total cut size (*i.e.*, the total weight of the hyperedges crossing the cut line) without violating a given area balance constraint. The area balance constraint prevents one partition from becoming much smaller than the other one. At any level of the recursion in the bisection approach, each region is considered in isolation from the other regions, although nets between regions are often modeled using *terminal propagation* [DK85]. In contrast, some placers extend bisectioning to multiway partitioning by allowing the intermediate results from the bipartitioning of a region to influence the final partitioning of additional regions at that level. Good examples of this approach include BONNPLACE [Vyg97] (discussed further in Section 5.4.1) and DRAGON [WYS00b] (discussed in more detail in Section 5.2.3), each of which uses recursive top-down quadrisectioning that divides each region being partitioned into four subregions, and FENGSHUI which uses k-way partitioning.

Compared to analytical and other placement approaches that target wirelength directly, partitioning-based placement is usually somewhat more effective at mitigating congestion. This is because the cut size across a local cut line has a stronger correlation to the local congestion in the vicinity of the cut line than does the total wirelength. Even so, it is still inadequate at capturing the full two-dimensional nature of congestion. Furthermore, pure partitioning-based placement is not used very widely in commercial applications because of placement stability concerns. Indeed, today's large design sizes and numerous macro blockages often require explicit congestion mitigation techniques even with partitioning-based placement approaches.

The major weakness of the top-down paradigm is that partitioning decisions at early stages of the recursion cannot be undone even if subsequent levels of the recursion indicate their undesirability. Thus, a cell that has been moved into the left half of the layout during the top-level partitioning will never be able to cross over to the right half later, even if all the cells connected to it end up in the far right of the region. This can lock the optimization down into local minima, resulting in increased wirelength.

5.1.3 Multilevel Placement Methods

Multilevel placement algorithms are a recent development [SR99, CCK+00] that remedies the main deficiency of pure top-down flows, which is the inability to recover from the selection of poor moves at early stages of the recursion. While they often rely on recent advances in algebraic multigrid methods, the basic intuition underlying these methods is quite simple.

At the abstract level, these algorithms involve three kinds of operations, namely, *coarsening*, *relaxation*, and *interpolation*. Coarsening refers to the process of building a hierarchy in a bottom-up manner (using recursive clustering) or top-down manner (using recursive partitioning). Relaxation refers to localized optimizations at each level of the hierarchy, while interpolation involves the transfer of an intermediate solution from one level of the hierarchy to the next. These operations can be organized in several different kinds of flows; the most commonly used flows involve one or more *V-cycles* in which the cells are first clustered together recursively to create a placement problem with fewer instances. Subsequently, the aggregated clusters are unclustered recursively, with local optimizations at each stage of the unclustering. These optimizations during the unclustering pass offer an opportunity to recover from poor choices made during the early stages of the clustering pass. This entire V-cycle can be repeated if necessary. So-called W-cycles (also referred to as backtracking V-cycles), in which the unclustering pass is interrupted after a few levels in order to redo the recursive clustering of those levels, followed by the recursive unclustering of all the levels, have also yielded good results. The multilevel placement paradigm has been discussed further in Section 5.3.1.

Many heuristics have been proposed to drive the clustering phase in multilevel placement. However, experiments indicate that simple graph-based greedy schemes such as First-Choice vertex matching [Kar99] may be more effective than sophisticated schemes that attempt to exploit some more extensive connectivity information. Another feature worth noting is that this paradigm merely provides a framework within which any kind of placement optimization can be applied at a given recursive level. Indeed, different multilevel placers have used analytical programming, simulated annealing as well as recursive partitioning in order to drive the relaxation at each level of their hierarchy.

In addition to the ability to recover from poor early decisions, multilevel placement approaches also offer the advantage of scalability. They can deal with large designs by merely adding additional recursive levels to their coarsening passes. At the same time, the optimizations applied at the coarsest levels of the hierarchy are good at exploiting the global structure of the netlist.

5.1.4 Move-based Methods

This class of placement techniques includes both simulated annealing as well as sequences of greedy moves. As discussed in Section 4.1.5 in Chapter 4, simulated annealing evaluates potential moves from any candidate solution. If a move improves the objective function, it is accepted. On the other hand, if it worsens the objective function, it still has a finite probability of acceptance given by $e^{-\Delta C/T}$, where ΔC is the increase in cost, and T is the so-called *temperature* parameter. This allows the optimization to escape from local minima within the solution space. As the placement proceeds, the temperature is reduced, thus decreasing the probability of accepting a move that worsens the cost function.

Simulated annealing can yield very high quality placements, but often requires very long runtimes to do so. It is the primary optimization engine within the TimberWolf placer [SS86]. Low temperature simulated annealing has been used for bin-swapping based refinement (in which entire blocks within a given recursive level are interchanged) at all but the finest levels of the Dragon [WYS00b] placer within a top-down recursive quadrisectioning framework. The multilevel placer mPG [CCP+03] (discussed further in Section 5.3.1) also uses simulated annealing for relaxation within each level of its clustering and unclustering passes.

Other move-based heuristics have also proven quite successful when combined with standard placement optimization techniques. These include greedy cell swaps as well as ripple-move sequences that yield local improvements to the layout at any stage. As an example, Dragon uses a detailed greedy strategy for cell swapping to refine the placement at its finest level. However, perhaps the most dramatic application of local moves has been within the quadratic programming framework employed by the FastPlace placer [CV04], in which each iteration of unconstrained quadratic wirelength minimization is followed by local cell shifts away from regions of high cell density and greedy cell swaps for wirelength reduction. This allows FastPlace to achieve extremely fast convergence of the layout without significant wirelength degradation.

5.2 Congestion-aware Post-processing of Placement

A given global placement can be post-processed to improve its congestion profile. The post-processing techniques used to achieve this can be broadly classified into three groups, namely,

- Find-and-fix methods (discussed in Section 5.2.1),
- Congestion-aware placement refinement methods (discussed in Section 5.2.2), and,

- White space management techniques (discussed in Section 5.2.3).

Find-and-fix congestion management techniques identify the locations of any congestion hot spots and then try to fix them using small local perturbations to the design without violating its convergence. The second class of post-processing techniques listed above relies on compute-intensive techniques to model the expected congestion directly into the objective function of the placement in addition to the traditional placement metrics, and then refine the given global placement further to yield a good, routable placement. Finally, the third class of post-processing techniques focuses on modifying the distribution of white space in the design in an attempt to make the design more routable.

5.2.1 Find-and-fix Techniques

The intuition behind this class of techniques is that the wirelength or cut size minimization objective of traditional placement does a decent job of reducing the average congestion, while generating a placement that is desirable with respect to the primary metrics of wirelength and/or delay. Its local congestion hot spots can be identified accurately and then improved effectively using only relatively minor perturbations that do not destroy the overall desirability of the placement. In contrast, incorporating inherently error-prone congestion estimation into the objective function that drives global placement may result in a local minimum with very poor wirelength or delay, even if it has improved congestion. Indeed, [WYS00a] studied various objective functions that captured different models for routing congestion or combined them with the wirelength within a simulated annealing based placer, and demonstrated that the use of pure wirelength as the objective function usually yielded a better starting placement even for subsequent congestion minimization than all their congestion-aware objective functions. However, the more sophisticated recent techniques for explicit congestion management during global placement (such as those for white space allocation) seem to be able to generate good global placements without significant deterioration of wirelength. But the congestion of even these placements can often be improved further using post-processing schemes.

The most widely used post-processing scheme employs simple trial-and-error; it first identifies the congestion hot spots in the initial placement based either on a probabilistic congestion map or by using a fast global router, and then makes some sequence of greedy local moves using the cells and/or nets lying within these congestion hot spots; a move is accepted only if it improves the congestion without deteriorating the other metrics significantly. Thus, one could attempt to swap a cell lying within a congestion hot spot with another cell that lies in some nearby uncongested region, as depicted in Fig. 5.2. A post-processing scheme of this kind is used to further improve the congestion-aware placement obtained using cell inflation proposed in [HYH+01], and is described along with that scheme in Section 5.4.1.

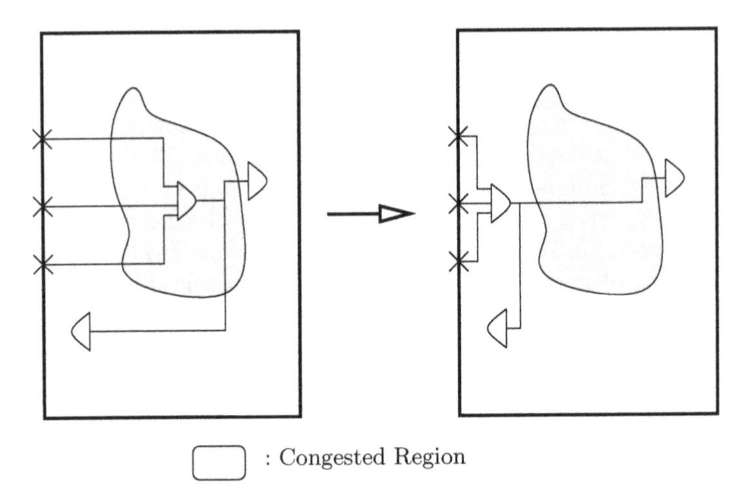

Fig. 5.2. Reducing congestion by moving cells lying within a congested region.

Alternatively, one could select the nets passing through highly congested regions, and sort them by their contribution to the total routing overflow within the design. Since any good global router will attempt to minimize the peak congestion by trying alternate routings for nets passing through congested regions (as discussed in Section 4.1 of Chapter 4), the mere rerouting of such nets during post-processing is usually not very effective at reducing the true congestion (although it can show benefits if the congestion profile had been estimated without actually running a global router). However, one can also move the cells that are connected to the pins of such nets, so that the routings of these nets no longer pass through highly congested regions. This is illustrated in Fig. 5.3.

The cell-centric scheme (that swaps out the cells lying in congested regions) focuses primarily on the congestion arising from local interconnections and the interconnects connected to local cells, while ignoring the congestion caused due to long global nets. In contrast, the net-centric scheme addresses all the sources of congestion. Three simple post-processing schemes have been studied in [WYS00a] within a simulated annealing based framework; these are variants of the greedy cell- and net-centric schemes described above, as well as a simple network flow based cell-centric approach that allows multiple cells to move simultaneously. Not surprisingly, they find that the net-centric approach yields the best results. A typical scheme based on this approach is presented in Algorithm 8.

The idea of post-processing a placement to improve its congestion characteristics has been around for quite a while. For instance, [TCT92] proposed the use of a congestion map based on a fast initial routing, in order to add a congestion cost to the objective function used to evaluate cell moves and pin assignments within a greedy iterative framework to post-process the place-

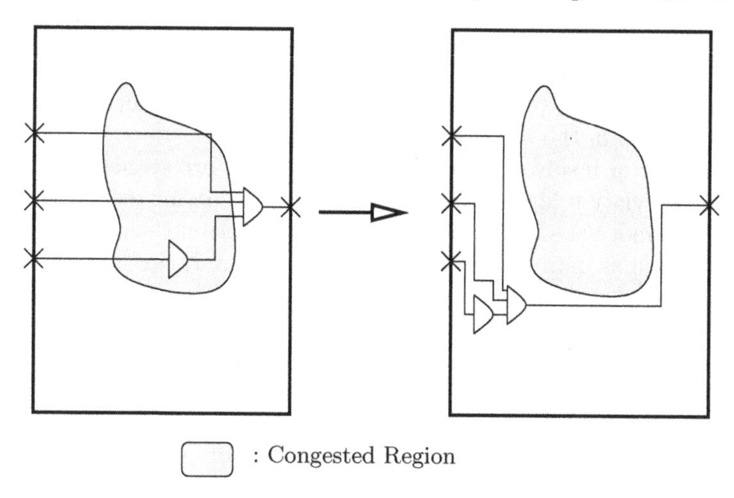

Fig. 5.3. Reducing congestion by moving cells connected to a net passing through a congested region.

Algorithm 8 Greedy net-centric post-processing for congestion relief

1: Sort list $\mathcal{N}$ of nets by decreasing contribution to total routing overflow
2: **for all** nets $n \in \mathcal{N}$ **do**
3: **for all** cells c connected to n **do**
4: Evaluate potential moves and swaps for c
5: Accept move that yields largest reduction in total overflow without violating timing constraints
6: **end for**
7: **end for**
8: Legalize placement

ment. In this approach, the congestion map is first generated from the initial placement using a forest of minimum wirelength spanning trees for the nets in the design (assuming that each of the two possible single bend embeddings of a diagonal edge in a spanning tree is equally likely), and then updated by iterating over the nets to choose congestion-aware embeddings for the edges in their spanning trees. Then, various swaps between pairs of cells and pairs of I/O pads, as well as their moves to unallocated cell slots and unused pad locations, are evaluated using a linear combination of wirelength, timing cost and congestion cost (by inspecting all the nets affected by any such move). The congestion map is updated whenever a move is accepted. Because of its greedy nature, the effectiveness of this approach depends on the order in which the nets are embedded during the congestion map generation and the order in which cell and pad moves are evaluated. It is also susceptible to getting trapped within local minima of the objective function. Over the years, variants of this approach have tried different heuristics to improve the performance of this approach by using more sophisticated ways to generate the initial congestion map, decide the order in which cell and pad moves are evaluated, and allow some moves that temporarily degrade the objective function or violate some constraints, in the hope of escaping local minima.

Another factor involved in designing a good post-processing algorithm for congestion alleviation is the extent of perturbation permitted in the design. Ideally, a cell move should be accepted only if it does not violate any design constraints such as path timing or clock skew. However, checking the validity of each such candidate move can become expensive. Therefore, an acceptable compromise can be the explicit checking of these constraints only for the cells that lie on the most critical paths (or prohibiting the movement of these cells completely), and merely bounding the extent of the moves for the remaining cells. However, if the bound on the extent of a move is too small, it may make the congestion alleviation problem unsolvable. There has been some work to help determine how large a region around a congestion hot spot should be permitted for cell moves [WYE+00, YWK+03]. In general, expanding each hot spot to a constant size or by a constant factor is usually too restrictive; it is better to allow somewhat larger expansion regions for the more serious hot spots. A good rule of thumb is that the expansion region for a hot spot should be somewhat larger than the smallest region such that the cumulative supply of routing tracks across all the global routing cells within this region exceeds the cumulative expected routing demand there. If the region is any smaller, the congestion within the hot spot cannot be resolved without excessive detours.

The order-dependency of the moves can be countered by using a more globalized formulation. For instance, an industrial network flow based formulation briefly described in [KRV02] identifies all the placement bins that have routing overflows, and sets up flow arcs from these bins to placement bins that have routing resources available. The solution of the resulting network flow problem instance moves cells from the congested bins to the sparsely routed bins, allowing a simultaneous determination of cell moves to reduce the total overflow. The flow arc costs and capacities within this formulation are set up such that the solution tries to maintain the relative logic ordering of cells as far as possible, in order to reduce the extent of the perturbation to the existing placement. This process of setting up network flow problem instances from the congestion map is iterated until all the routing overflows have been eliminated, or the iterations yield no further benefits. This formulation relies on the correlation between cell density and the local routing congestion. Consequently, it does not resolve congestion problems arising due to non-local flyover routes.

For many years, find-and-fix post-processing schemes were used widely to improve the routability of placements generated by commercial tools. However, the wiring complexity of modern designs requires that such schemes be augmented by congestion awareness in upstream placement optimizations, using, for instance, techniques described in Sections 5.3 and 5.4, because of insufficient flexibility available if the global placement has already been completed.

5.2.2 Congestion-aware Placement Refinement

This class of techniques relies on the direct modeling of the expected congestion within the objective function to be optimized during placement. In that sense, it is similar to the techniques described in Section 5.4. However, it differs from those techniques in that it typically uses compute-intensive congestion models or placement paradigms that make it unsuitable for use as a primary engine for global placement.

Modeling Local Unroutability using Net Weights

A good example of a congestion-aware placement refinement technique is the SPARSE scheme presented in [HM02], in which a simulated annealing based placer is used with an objective function consisting of the usual weighted sum of expected netlengths, but with the weights dynamically capturing the routability of the corresponding nets. This work is motivated by the authors' empirical observation that around 75% of the routing resources within a global placement bin are used up by nets that have a pin within that bin[2]. Thus, moving the pins of nets out of bins that are likely to be congested and into bins that do not show congestion problems usually improves the overall routability of the design. Furthermore, it also usually helps with good white space allocation, thus allowing the placement to be legalized more easily.

Given a design with nets $\{n_i\}_{i=1,2,\ldots,N}$ and a traditional placement objective function $\sum_i w_i l_i$, where w_i is the weight associated with net n_i (that is often used to model the timing criticality of the net), and l_i is its expected length (obtained using the same fast estimates that are used during global placement), SPARSE modifies the net weight w_i to also reflect the predicted routability problems in the bins containing the pins of the net n_i. It does so by

[2] Although this percentage may be somewhat lower for large industrial designs, a significant proportion of the routability problems in a design can nevertheless be root-caused to pin hookup issues. This is not surprising when one considers that a net having a pin within a given bin requires a via stack to connect that pin to the metal layer on which the net has been routed; this via stack creates a blockage on all the layers lying between the pin and the route of the net. Furthermore, the local switchbox routing carried out by detailed routers for the purpose of route completion also often creates numerous small metal segments and vias in the route of a net in the vicinity of its pins. In contrast, a net routed through a bin that does not contain any of its pins often uses up merely a single routing track on a single layer within that bin. Even if the route of the net has a Steiner node involving multiple layers lying within the bin, the corresponding vias often involve only the upper layers (whose routability is usually less degraded due to short metal segments and via stack blockages than the lower layers).

replacing each net weight w_i with the corresponding weight $w_i \frac{\sum_B p_B}{d_i}$, where the summation is taken over all the global placement bins B that contain some pin of n_i, the number of such bins is d_i, and p_B is a "congestion parameter" for B. As shown in the right-hand side of Fig. 5.4, the congestion parameter for a bin attempts to capture the likelihood of that bin having routability problems. It does so by first computing a proxy D_B for the expected routing demand for that bin (as described in the next paragraph), and then transforming that demand into a smooth approximation of a threshold function modeling the routing overflow, via a fitted exponential function $p_B = \alpha D_B^{\beta} + \gamma$ (where α, β and γ are the parameters used for fitting). This exponential function captures the intuition that small values of D_B are not likely to cause routability problems; however, as D_B grows beyond some threshold that corresponds to the number of routing tracks available within the bin, the likelihood of the corresponding bin having a routing overflow grows very rapidly. Given this intuition behind p_B, the $\frac{\sum_b p_B}{d_i}$ multiplier for the weight w_i corresponding to net n_i captures the average congestion parameter for all the bins that contain its pins, creating a strong incentive for the placer to move cells out of bins that are likely to be congested.

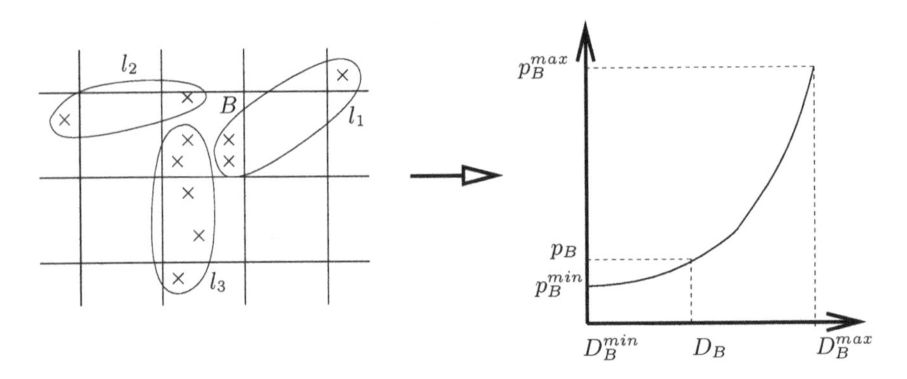

Fig. 5.4. Computing the congestion metric p_B for a bin B as in [HM02].

In order to compute the proxy D_B for the demand of a bin, the estimated wirelength l_j of each net n_j is distributed equally among all its pins, so that the share of each pin is $l_j/|n_j|$, where $|n_j|$ is the number of pins in n_j. Then, for each bin B, D_B is merely the sum of the (weighted) netlengths associated with the pins lying in B (*i.e.*, computed as $\sum_j \frac{w_j l_j}{|n_j|}$, where the summation is carried out over all nets that have a pin in B).

An example of the computation of the congestion parameter p_B for a bin B is illustrated in Fig. 5.4. In this figure, let the estimated wirelengths for the three nets with pins in B be l_1, l_2 and l_3, respectively. Then, p_B is given by

$\alpha D_B^\beta + \gamma$, where $D_B = \frac{l_1}{3} + \frac{l_2}{2} + \frac{l_3}{5}$ (assuming $w_i = 1$, for $i = 1, 2, 3$), since the three nets consist of three, two and five pins, respectively.

Interleaving of Cells and Wire Segments within a Row

Another sophisticated placement refinement approach presented in literature [JL04] incorporates the assignment of wire segments into a dynamic programming based detailed placement framework that allows the optimal interleaving of two sets of cells within a row of the layout [HL00]. In contrast to greedy approaches, this approach allows the optimal search of an exponential sized subset of the complete solution space within polynomial time. Although this approach has been experimentally verified only in the context of field-programmable gate arrays (FPGAs) and is limited to some extent by the upfront need to partition the cells and wire segments within a row into two sequences for interleaving, the underlying ideas are novel and promising, and are also applicable to the detailed placement of (row-based) standard cell designs.

Since this approach builds upon the MONGREL algorithm for relaxation-based local search [HL00], it is useful to briefly review how MONGREL uses *interleaving* for wirelength-driven detailed placement, before describing its extensions to improve routability. Given a partitioning of the cells in a row into two disjoint sequences $A = (a_1, \ldots, a_m)$ and $B = (b_1, \ldots, b_n)$, such that, for any valid i, the ith cell within a sequence precedes the $(i+1)^{th}$ cell within that sequence in the given placement of the row, interleaving explores all possible permutations for these cells that preserve the relative orders within each of these two sequences. Thus, if a row is partitioned into A containing four cells and B containing five cells, a permutation $a_1a_2b_1a_3b_2b_3b_4b_5a_4$ is considered valid, while another permutation $a_1a_3b_1a_2b_2b_3b_4b_5a_4$ is not (because a_3 and a_2 are reversed in the latter permutation). Each of the valid permutations is evaluated for its corresponding wirelength. Let $S_{i,j}$ denote the optimal interleaving of the sequences $a_1, \ldots, a_i$ and $b_1, \ldots, b_j$, and $C(S_{i,j})$ denote the cost of this interleaving. Then,

$$S_{0,0} = \emptyset,$$

$$C(S_{0,0}) = 0,$$

$$C(S_{i,j}) = \min\{C(S_{i-1,j}a_i), C(S_{i,j-1}b_j)\}.$$

Stated simply, this recurrence states that the best interleaving of the first i cells in A and the first j cells in B is obtained by appending one of a_i and b_j to the best interleaving of the remaining $i+j-1$ cells, with the better option between a_i or b_j being selected.

Given this background for MONGREL, [JL04] extends it to include the wire segments crossing the row vertically into the interleaving formulation. While some vertical segments (such as s_1 in Fig. 5.5) terminate at a cell and hence

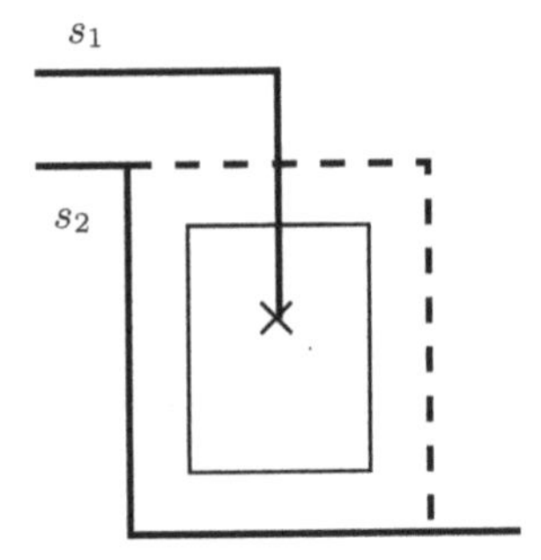

Fig. 5.5. Assignment of a free vertical net segment to either side of a cell can have different costs.

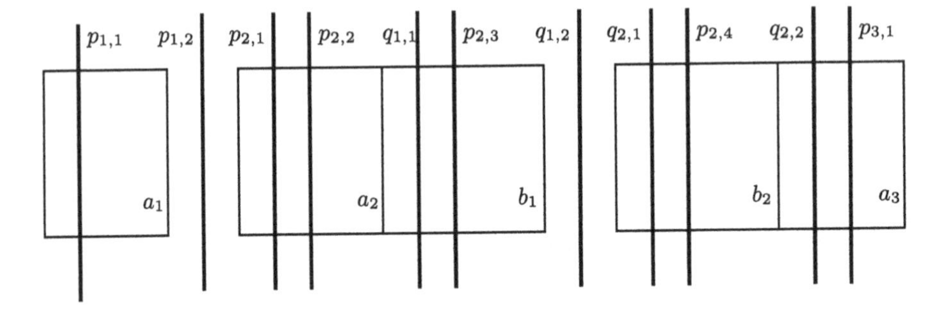

Fig. 5.6. An example of an interleaving of some cells and vertical wire segments.

must move along with that cell whenever it is moved, others such as s_2 in the same figure have the flexibility of being assigned to either side of any given cell (with differing congestion and possibly differing wirelength costs). Given an initial assignment of cells and vertical segments to specific locations along a row, let the cells be partitioned into two sequences A and B as before. Let the vertical segments also be partitioned into two disjoint sequences $P_1 \cup \ldots \cup P_m$ and $Q_1 \cup \ldots \cup Q_n$, where $P_i = \{p_{i,1}, \ldots, p_{i,|P_i|}\}$ and $Q_j = \{q_{j,1}, \ldots, q_{j,|Q_j|}\}$, such that all segments belonging to P_i (respectively, Q_j) precede any segment in P_{i+1} (respectively, Q_{j+1}). Furthermore, let all the segments in P_i be routed over or to the right of cell a_i but to the left of a_{i+1}, and all the segments in Q_j be routed over or to the right of cell b_j but to the left of b_{j+1}. Figure 5.6 presents an illustration of such a partitioning and of the notation introduced here. Finally, let $S_{i,j,k,l}$ represent the sequence in which all cells $a_1, \ldots, a_i$ and $b_1, , \ldots, b_j$ as well as vertical segments belonging to $P_1 \cup \ldots P_{i-1}$ and $Q_1 \cup \ldots Q_{j-1}$ have been placed in an optimal interleaving, along with the first k segments from P_i and the first l segments from Q_j. Let $C(S_{i,j,k,l})$ be the cost of this interleaving. Then, the recurrence for the cost of an interleaving is given by:

Algorithm 9 Optimal congestion-aware interleaving within the MONGREL framework

1: Given sequences $A : (a_i, P_i)$ with $0 < i \leq m$ and $B : (b_j, Q_j)$ with $0 < j \leq n$,
2: $S_{0,0,0,0} \leftarrow \emptyset$
3: $C(S_{0,0,0,0}) \leftarrow 0$
4: **for all** $i = 1, \ldots, m$ **do**
5:   **for all** $j = 1, \ldots, n$ **do**
6:     **for all** $k = 0, \ldots, |P_i|$ **do**
7:       **for all** $l = 0, \ldots, |Q_j|$ **do**
8:         $C_A \leftarrow \infty$
9:         **for all** $x = 0, \ldots, l$ **do**
10:           **if** $C(S_{i-1,j,|P_{i-1}|,x}) + C(a_i) + C(p_{i,1}, \ldots, p_{i,k}) + C(q_{j,x+1}, \ldots, q_{j,l}) < C_A$ **then**
11:             $C_A \leftarrow C(S_{i-1,j,|P_{i-1}|,x}) + C(a_i) + C(p_{i,1}, \ldots, p_{i,k}) + C(q_{j,x+1}, \ldots, q_{j,l})$
12:           **end if**
13:         **end for**
14:         $C_B \leftarrow \infty$
15:         **for all** $y = 0, \ldots, k$ **do**
16:           **if** $C(S_{i,j-1,y,|Q_{j-1}|}) + C(b_j) + C(p_{i,y+1}, \ldots, p_{y,k}) + C(q_{j,1}, \ldots, q_{j,l}) < C_B$ **then**
17:             $C_B \leftarrow C(S_{i,j-1,y,|Q_{j-1}|}) + C(b_j) + C(p_{i,y+1}, \ldots, p_{y,k}) + C(q_{j,1}, \ldots, q_{j,l})$
18:           **end if**
19:         **end for**
20:         $C(S_{i,j,k,l}) \leftarrow \min\{C_A, C_B\}$
21:         Record which of C_A or C_B is selected for $S_{i,j,k,l}$, along with corresponding value of x or y
22:       **end for**
23:     **end for**
24:   **end for**
25: **end for**
26: Recover optimal interleaving $S_{m,n,|P_m|,|Q_n|}$ by backtracing the subsequences used to derive the optimal cost (by following the recorded choices of C_A or C_B and corresponding values of x or y at each stage)

$$S_{0,0,0,0} = \emptyset,$$

$$C(S_{0,0,0,0}) = 0,$$

$$C(S_{i,j,k,l}) = \min\{C_A, C_B\},$$

where,

$$C_A = \min_{0 \leq x \leq l} \{C(S_{i-1,j,|P_{i-1}|,x}) + C(a_i) + C(p_{i,1}, \ldots, p_{i,k}) + C(q_{j,x+1}, \ldots, q_{j,l})\},$$

$$C_B = \min_{0 \leq y \leq k} \{C(S_{i,j-1,y,|Q_{j-1}|}) + C(b_j) + C(p_{i,y+1}, \ldots, p_{y,k}) + C(q_{j,1}, \ldots, q_{j,l})\}.$$

In words, $C(S_{i,j,k,l})$ is obtained by taking the best interleaving for the first $i+j-1$ cells as in MONGREL, but now it also includes the assignment for all the vertical segments that precede the last cell in the interleaving for the first $i+j-1$ cells. Then, the best possible way of arranging the remaining segments around the last cell in the interleaving is selected. For instance, if the last cell in the interleaving is a_i, then all the segments in P_i must lie either over it or to its right. However, there is no restriction on where the segments in Q_j lie in relation to this cell, as long as they obey their ordering within Q_j. This is exploited in the computation of C_A above, by splitting this sequence on either side of a_i in the best possible way, by keeping the first x wire segments to its left (for the best possible x). The computation of C_B is analogous. This process is summarized in Algorithm 9.

While this framework does not guarantee the absolute optimum for the congestion cost over all possible configurations of placement and local routing, it does explore a significant fraction of the solution space optimally in a very efficient manner. Furthermore, it is targeted towards wirelength and routability improvement through local perturbations during the detailed placement stage after a pass of global routing has been carried out, resulting in an accurate estimation of localized congestion costs and wirelengths.

5.2.3 White Space Management Techniques

The *white space* within a design block is defined as the area within the block that is not occupied by cells in its final layout. In recent years, there has been a surge of research interest in techniques to manage white space effectively during placement. While some of this work is motivated by performance or wirelength considerations, much of it targets the improvement of the routability of the design. Some of the proposed white space management techniques are applicable subsequent to the global placement and are discussed in this section, while others operate concurrently with the global placement and are described in Section 5.4.2. As with the SPARSE [HM02] scheme discussed earlier in Section 5.2.2, increasing the white space in a congested region helps if the congestion is caused primarily due to pin accessibility problems for the nets having at least one pin within that region.

Row-based White Space Allocation

The work presented in [YCS03] presents an extensive study of several white space allocation schemes targeting routability improvements within the context of the placer DRAGON [WYS00b] that uses recursive quadrisectioning interleaved with simulated annealing. This work includes the comparison of various schemes for allocating white space to each placement bin at the end of global placement, based on its congestion as measured by its estimated routing overflow (with respect to some threshold). In particular, it studies variants of direct schemes that allocate white space at the bin level, and compares them against variants of two-step schemes that first distribute white space among the rows, and then within each row. These variants of direct and two-level schemes include allocation schemes that are based on different thresholds for overflow computation, as well as allocation schemes that use different (*i.e.*, linear or quadratic) functions of the congestion to determine the extent of white space allocation to a placement bin.

Direct schemes that operate at the bin level cannot control the amount of white space available within a row. Consequently, the design can end up with some rows that have no white space at all. Since the allocation of white space usually causes some degradation in the primary placement metrics (*i.e.*, wirelength or delay), the white space allocation phase must usually be followed by some local improvement of the placement in order to recover from as much of the placement degradation as possible. However, a post-allocation modification to a row that has no white space available can often cause a row length violation, whose correction can result in a large perturbation to the design. Therefore, two level schemes that ensure a minimum amount of white space for each row are usually more amenable to good quality placements that are also routable, than direct schemes.

Even among two-level schemes, one can distinguish between schemes that ensure a minimum bound on the white space within each row, and those that ensure both minimum and maximum bounds. While a lower white space bound for a row helps the row to absorb post-allocation optimization perturbations, an upper bound is useful in ensuring that excessive white space within a single row does not cause severe wirelength degradation. However, the optimal values for these bounds vary from design to design, depending on both the total available white space and the severity of congestion in the design.

Another concern in designing a good white space allocation scheme relates to the aggressiveness of the allocation, as depicted in Fig. 5.7. For the same distribution of congestion among a given set of bins, an aggressive scheme will allocate increased white space to the more congested bins at the cost of white space allocated to the less congested ones. An allocation scheme can be made aggressive by, say, using a quadratic function of the congestion of a bin to determine the amount of white space to be assigned to it, instead of a linear function. Alternately, an aggressive scheme can define the congestion as the overflow with respect to the average congestion across the design, instead of the minimum congestion. In general, aggressive schemes have a greater positive impact on the routability of a design (especially when the total available white space is limited), but can also cause greater degradation of the primary placement metrics. Thus, they are appropriate for highly congested, densely placed designs, but degrade the placement quality without yielding a commensurate routability benefit on sparse designs that are not very congested.

Let W be the total white space in the design, with n being the number of rows in the design. Let the congestion of row j (as measured by the total

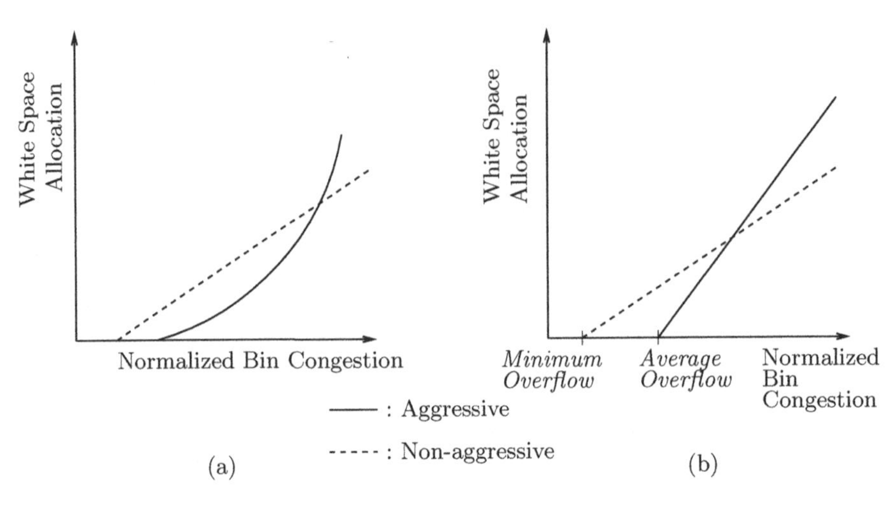

Fig. 5.7. Aggressiveness of white space allocation strategy: (a) Linear versus quadratic dependence on congestion. (b) Use of minimum overflow versus average overflow to decide white space allocation threshold (Linear curves are shown; quadratic curves will also be similar).

overflow of the cells in the row with respect to some specified threshold) and the white space to be allocated to row j be denoted by c_j and w_j, respectively (so that $\sum_{i=1}^{n} w_i = W$). Then, if $w_{\min}$ is the minimum white space to be allocated to a row, a linear allocation of the white space yields:

$$w_j = w_{\min} + \frac{W - nw_{\min}}{\sum_{i=1}^{n} c_i} c_j.$$

Next, the white space allocated to a row is distributed among its bins. A linear allocation here yields:

$$w_{ij} = \frac{w_j c_{ij}}{c_j},$$

where w_{ij} is the white space allocated to the ith bin (with congestion c_{ij}) belonging to row j (with $\sum_k c_{kj} = c_j$). Similarly, a quadratic allocation yields:

$$w_{ij} = \frac{w_j c_{ij}^2}{\sum_k c_{kj}^2}.$$

If the two-level row-based allocation is to also obey upper bounds (say, $w_{\max}$) on the white space allocated to each row, the allocation function must assign $w_{\min}$ units of white space to the least congested row (with congestion being, say, c_1) and $w_{\max}$ units of white space to the most congested row (with congestion being, say, c_n), while still ensuring that the total white space allocated across all the rows is W. All these constraints cannot be simultaneously guaranteed by a linear allocation function, but a quadratic allocation function:

$$w_j = a_1 c_j^2 + a_2 c_j + a_3$$

suffices. Here, a_1, a_2, and a_3 are coefficients whose values can be obtained by solving the three constraints mentioned above for $w_{\min}$, $w_{\max}$ and W. However, if $c_1 < -\frac{a_2}{2a_1} < c_n$, the allocation function results in the extremum lying inside $[c_1, c_n]$ (so that it is not monotone within this range). In order to make the function monotone again, one of the $w_{\min}$ or $w_{\max}$ constraints must be relaxed. In this case, if $a_1 < 0$, then the $w_{\min}$ constraint is relaxed and the point $(c_n, w_{\max})$ is set as the extremum of the quadratic function by setting $c_n = -\frac{a_2}{2a_1}$. On the other hand, if $a_1 > 0$, then the $w_{\max}$ constraint is relaxed and the point $(c_1, w_{\min})$ is set as the extremum of the quadratic function by setting $c_1 = -\frac{a_2}{2a_1}$. Once the row white space allocation has been completed using this quadratic function, the white space assigned to each row can be distributed among its bins using a linear or quadratic function as described previously.

Algorithm 10 Row-based white space allocation within DRAGON

1: **while** global placement bin size not small enough **do**
2: Carry out a few iterations of recursive partitioning of the circuit and geometric slicing of the layout area
3: Place the resulting clusters in their corresponding bins
4: Use wirelength-driven low temperature simulated annealing to improve the clusters
5: **end while**
6: Adjust the global placement bins to match the row structure of the design
7: Estimate the routing overflow in each bin
8: Perform two-level row-based white space allocation
9: Carry out a wirelength-driven low temperature simulated annealing based local optimization without degrading white space in each bin
10: Estimate the routing overflow in each bin
11: Perform two-level row-based white space allocation
12: Legalize the placement
13: Carry out a wirelength-driven low temperature simulated annealing based local optimization without degrading white space in each bin or creating overlaps

The work described in [YCS03] proposes to use the row-based white space allocation twice as a post-processing step after the DRAGON global placement, as summarized in Algorithm 10. More specifically, the global placement flow consists of a top-down recursive partitioning phase during which the design is partitioned into clusters that are placed into bins, interleaved with a low temperature simulated annealing of the clusters that is driven by wirelength. Once the clusters (and the corresponding bin sizes) are small enough, the detailed placer first adjusts the bins to match them to the underlying row structure of the design. Next, a wirelength-driven simulated annealing phase is followed by legalization and local improvements. The white space allocation step intercepts the default DRAGON flow first after the bin adjustment, and then again just before the legalization phase. Each of these white space allocation phases is followed by a low temperature simulated annealing phase involving cell moves and swaps to improve the wirelength without changing the white space allocation of each bin significantly. Overall, the congestion-aware version of DRAGON has proven to be quite effective at producing routable layouts with good wirelengths. While the placement runtime for this algorithm is often significantly higher than several other competing approaches, it is usually adequately compensated by considerable runtime improvements during the routing phase due to the routing problem generated by the placement being easier, resulting in an overall reduction in the total layout runtime.

Cut Line Adjustment

The white space allocation scheme presented in [LXK+04] is a post-processing scheme that is applicable to any global or detailed placement produced using any arbitrary placement engine. The primary operation within this scheme is that of cut line adjustment. Given a placement of a design, this scheme recursively partitions the layout based on the geometric locations of its cells. Each bisection divides a partition into two equal sized partitions. This sequence of recursive partitions can be represented by a binary slicing tree, as shown in Fig. 5.8. In this tree, the root represents the top-level bisection cut. At any level within the tree, the two child nodes of a node represent the two partitions created by the corresponding cut. The leaf nodes of the tree represent the final set of partitions obtained by the partitioning of the given placement.

Next, a bottom-up congestion analysis is carried out for each node of the slicing tree. For a leaf node, the congestion is merely the total routing overflow for the global routing bins corresponding to that node, while that for an internal tree node is the sum of the congestion levels of its child nodes. Once the congestion values have been estimated for each node in the slicing tree, the cut lines corresponding to each node in the tree are adjusted on the layout in a top-down fashion so as to make the amount of white space available to the child nodes of a node linearly proportional to their congestion levels. Consider a region with lower left corner at (x_{ll}, y_{ll}) and upper right corner at (x_{ur}, y_{ur}) (so that its total area $A = (x_{ur} - x_{ll})(y_{ur} - y_{ll})$) and total cell area C (so that its white space is $A - C$). If a vertical cut line originally at $(x_{ll} + x_{ur})/2$ divides the region into a left subregion with congestion level χ_0 and cell area C_0, and a right subregion with congestion level χ_1 and cell area C_1 (with $C_0 + C_1 = C$), the redistributed white space allocated to the two subregions is $(A - C)\frac{\chi_0}{\chi_0+\chi_1}$ and $(A - C)\frac{\chi_1}{\chi_0+\chi_1}$, respectively. Consequently, the updated location of the cut line is given by:

$$x_{cut} = \gamma x_{ur} + (1 - \gamma)x_{ll},$$

where,

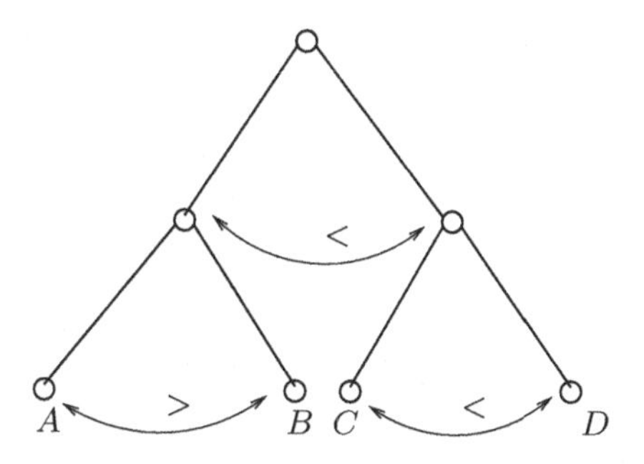

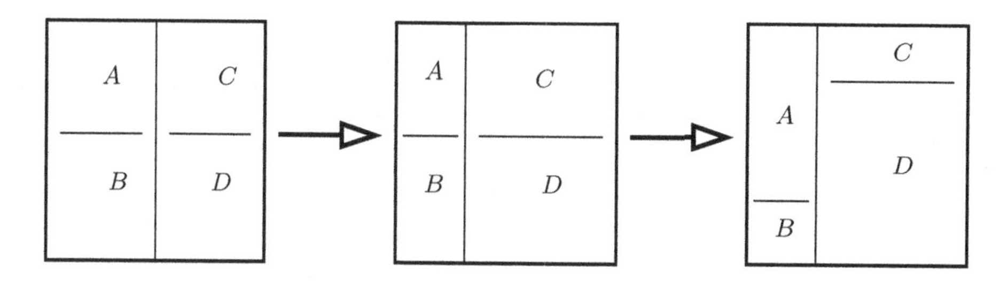

Fig. 5.8. White space allocation through hierarchical cut line adjustment.

$$\gamma = \frac{C_0 + (A - C)\frac{\chi_0}{\chi_0 + \chi_1}}{A}$$

is the ratio of the area of the left subregion to that of the original region after the cut line adjustment. The location of horizontal cuts is computed in a similar manner. The cells within each partition are constrained so that their center of gravity is the updated center of the partition after the cut line adjustment. The top-down cut line adjustment phase is followed by detailed placement and legalization.

Thus, in the example depicted in Fig. 5.8, since the total congestion for the A and B partitions is less than that of C and D (as indicated by the "<" between the corresponding nodes in the figure), the top-level cut line is shifted to the left in a way that distributes the total available white space proportional to these congestion levels. The same process is repeated at the next level of the partitioning, with the cut line between A and B being shifted downwards, and that between C and D being shifted upwards. In this manner, the white space available within a placement is redistributed so that regions with routing problems are allocated more white space in the hope of easing the congestion, at the cost of the white space in sparse regions that currently have no routability issues.

5.3 Interleaved Congestion Management and Placement

Several of the find-and-fix and white space allocation post-processing techniques for congestion management do not rely on the existence of a completely placed design, and can be adapted to work even while the global placement is still evolving. Since most of the techniques for global placement are iterative, they can be intercepted after every few iterations in order to perturb the approximate placement obtained so far to improve its congestion characteristics. This improvement is obtained by first identifying the congestion hot spots in the approximate placement by either using some congestion model (such as a probabilistic estimate of the routing) or by running a fast global router. Next, these hot spots are improved by either moving some cells explicitly or by influencing the metrics (such as net weights) that will drive the next few iterations of the placer (using techniques such as those described in Section 5.4), in a way that tends to reduce the congestion problems. Thus, compared to pure post-processing approaches, potential congestion problems are identified earlier and can be resolved while there is still sufficient flexibility available in the evolving placement.

However, there are two potential problems with the interleaved approach that a good congestion management flow must guard against. Firstly, during the early stages of the global placement, the location of a cell is very approximate, so that congestion hot spots may not be identified correctly or the cell moves or metric modifications that are expected to resolve these hot spots may not be very effective. Instead, these fixes may end up constraining the placer in the optimization of its primary objective function (namely, wirelength or timing) without getting any congestion benefits in return. This problem can be alleviated to some extent by intercepting the placement iterations only after the global placement has stabilized to some extent, and increasing the frequency of these interceptions as the placement converges.

The second potential risk with this approach involves the tradeoff between runtime overhead due to the interceptions and the extent of perturbation caused due to each interception, that may make convergence more difficult. If successive interceptions are separated by too many iterations, the placement may have changed sufficiently from the last interception so that a large number of new cell moves or metric modifications are required, thus jeopardizing the convergence of the placement. On the other hand, if the interceptions are too close to each other (so that the total number of interceptions is large), their total runtime overhead may be prohibitive. To some extent, this problem can be ameliorated by using low overhead congestion models explicitly within the placement process itself without requiring explicit interleaving with congestion estimation and optimization modules (as described in Section 5.4), although the challenge here is usually the accuracy of these congestion models and fidelity of the fixes. Furthermore, even a significant runtime overhead is often acceptable during placement if it leads to an easier routing problem (since the total design implementation time for complex, congested designs is often dominated by the router).

5.3.1 Interleaved Placement and Global Routing

One of the earliest works to introduce the idea of interleaving the placement with the routing was [SK89], using quadrisection-based placement. In this approach, routing topologies are generated for all the newly cut nets after each quadrisection of a design block; each topology spans the current set of design blocks that contain one or more terminals for the corresponding net. This allows the topology of a net to evolve hierarchically as the quadrisection progresses, as depicted in Fig. 5.9, ending with a complete tree spanning all the cells connected to that tree. More significantly, the specification of the spanning trees for the nets at any stage in the quadrisection process influences the subsequent placement decisions by embedding the tree edges into specific cuts of the partition (thus tracking the actual routing resource utilization along each cut, and avoiding a spurious routing resource penalty for a net on cuts between adjacent blocks containing pins of that net but not connected directly to each other, such as the blocks containing terminals a and c in the figure).

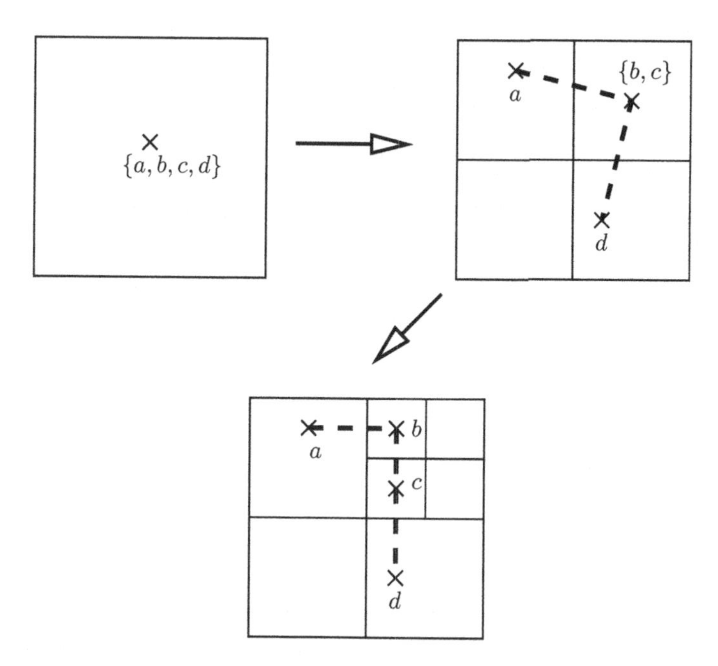

Fig. 5.9. Interleaved partitioning-based placement and interconnect topology generation.

However, the overall approach of [SK89] suffers from an inability to recover from poor early choices in cell partitioning or tree topology generation, potentially leading to poor layout because of inaccurate early estimates for the metrics of interest (such as wirelength or timing). Furthermore, for the congestion estimates to be consistent with the final routing, the interconnect trees generated during the placement are not allowed to be subsequently rerouted during the route completion phase. This can lead to severe problems in route completion in congested designs (since the rip-up and reroute technique is one of the most powerful route completion methods available to routers, and preventing its use can destroy the effectiveness of a router). Thus, while fixing the route topology for a small number of critical nets during early placement can be an effective way of making them predictable (although not of making the congestion estimates more accurate), doing so for a significant fraction of nets often leads to an infeasible routing problem.

Partitioning-based Quadratic Placement

Recently, [PBS98] has presented a scheme to integrate global routing with quadratic placement in order to make the resulting placement more routable. Their scheme is incorporated into the GORDIAN quadratic placement algorithm [KSJ+91] discussed in Section 5.1.1. The scheme proposed in [PBS98] intercepts GORDIAN after each partitioning iteration to run a fast global router to estimate the routing congestion within each partition, and then adjusts the area of each partition accordingly, as outlined in Algorithm 11. The routing congestion estimate considers both inter-partition routes computed using a region router operating at the granularity of the partitions, as well as intra-partition routes computed using a fast single trunk Steiner tree heuristic for multipin nets and single bend ("L") routes for two-pin nets. (Thus, in contrast to the inter-partition congestion that is computed using an actual router, the intra-partition congestion is an approximation based on fast congestion-oblivious topologies for the local nets). The total routing demand within a partition is then compared against the routing resources available within that partition in order to determine the expected congestion. This is then translated to a partition weight:

$$w_j = 1 - \frac{1}{\Delta_j}(s_j - d_j)$$

for each partition P_j, where s_j, d_j, and Δ_j are, respectively, the routing supply, routing demand, and size of the partition along the dimension (*i.e.*, x or y) being optimized in the current iteration, expressed in terms of routing tracks. This adjusts the weights of the partitions linearly around 1, with congested partitions getting larger weights, and all weights being positive (since $s_j \leq \Delta_j$ and $d_j \geq 0$). These weights are then embedded into a diagonal matrix G whose entry g_{ii} equals the partition weight of the independent cell c_i. If we let $\mathbf{p}'_{\mathbf{I}} = G\mathbf{p}_{\mathbf{I}}$ and substitute for $\mathbf{p}_{\mathbf{I}}$ in the expression for Φ_I in Equation (5.4), we find that Φ_I is minimized by solving:

$$(G^{-1}Z)^T C(ZG^{-1})\mathbf{p}_{\mathbf{I}} + \mathbf{d}'_{\mathbf{I}} = \mathbf{0},$$

where,

$$\mathbf{d}'^T_{\mathbf{I}} = (C\mathbf{p}^{(0)} + \mathbf{d})^T G^{-1} Z.$$

Just as with GORDIAN, this is guaranteed to be a positive definite system. Solving for Φ_I automatically scales the cell locations by their corresponding partition weights, thus expanding congested regions and shrinking regions with sparse routing. More specifically, if a region appears to be, say, vertically congested (*i.e.*, if it has an insufficient number of horizontal tracks), a vertical expansion or horizontal compression is performed, causing some of the internal horizontal nets to become vertical and thus relieving some of the vertical congestion.

Algorithm 11 Congestion relief through routing interleaved with quadratic placement

1: **while** placement not converged **do**
2: Carry out quadratic placement iteration
3: Estimate inter-partition routing using region-based router
4: **for all** partitions P_j in current partitioning level **do**
5: Compute congestion-oblivious internal routing demand for P_j using single trunk Steiner heuristic for multipin nets
6: Estimate partition weight for P_j based on internal and external routing demands and available routing resources within P_j
7: **end for**
8: Set up partition weight diagonal matrix G
9: Compute new system matrix $(G^{-1}Z)^T C(ZG^{-1})$ and vector $(C\mathbf{p}^{(0)} + \mathbf{d})^T G^{-1} Z$
10: **end while**

While GORDIAN has been superceded by more sophisticated analytical placement based approaches such as KRAFTWERK [EJ98] or APLACE [NDS01, KW04], the underlying idea proposed in [PBS98] of intercepting the iterations of the analytical placer periodically to estimate the routing congestion and then spreading out the congested regions is still an effective way of reducing routing congestion.

Multilevel Placement

Another recent example of interleaving placement and routing was presented in [CCP+03] in the context of a V-shaped multilevel framework for global placement. As shown in Fig. 5.10 and discussed in Section 5.1.3, this framework first includes a coarsening phase in which node clustering is used to recursively build coarser levels until the netlist is small enough to be placed efficiently. This initial placement is followed by a refinement phase in which each coarse level placement is unclustered recursively to obtain a finer level placement. This framework is independent of the choice of placement paradigm used for the initial placement at the coarsest level or for the placement refinement at each refinement level (although [CCP+03] presents data in the context of an implementation called MPG that uses simulated annealing).

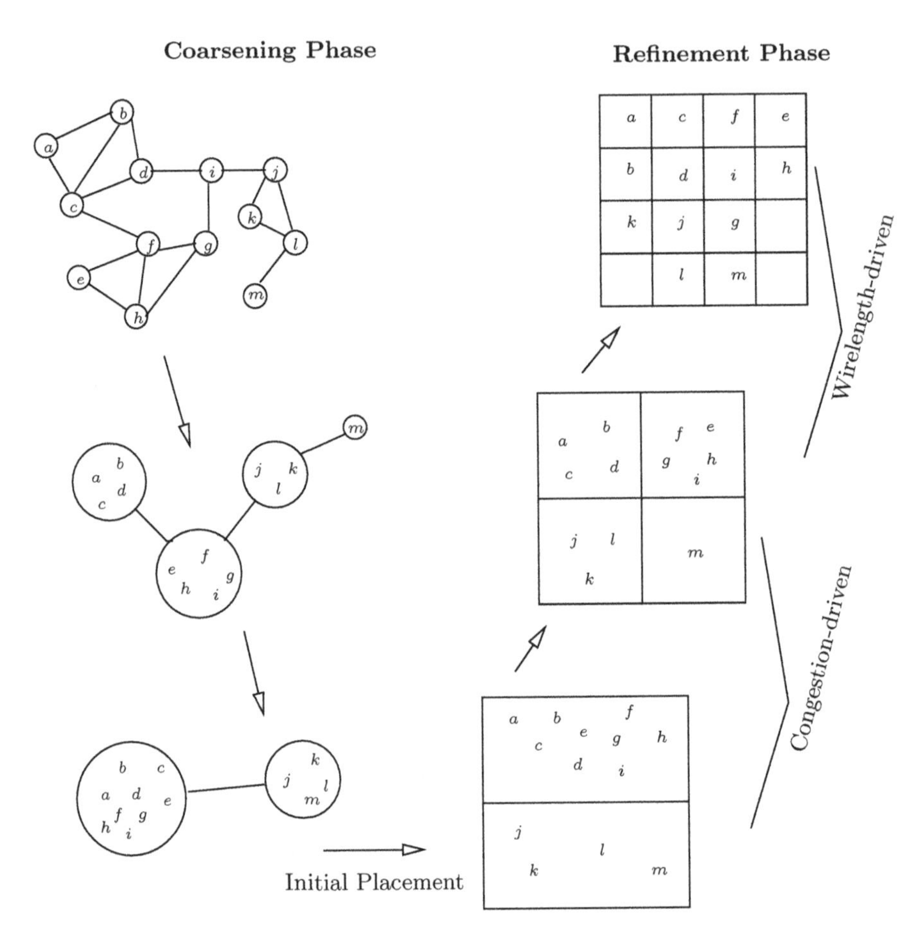

Fig. 5.10. Outline of a congestion-aware multilevel global placement framework.

The multilevel framework is applied to the congestion-aware placement in [CCP+03] by interleaving the placement with a very fast incremental global router. This interleaving is carried out at a very fine granularity, namely, that of each placement move within the simulated annealing engine. The routing is performed using a congestion-aware two-bend ("LZ") router for two-pin nets, along with a fast incremental Steiner arborescence tree (A-tree) generator for multipin nets. Periodically, layer assigment of the nets is also carried out based on their criticality, with critical nets being assigned to higher layers.

At the coarse levels of the multilevel framework, the cost function for the placement is merely the sum of the bounding boxes of the nets, because of the difficulty of estimating the congestion accurately at these coarse levels. In contrast, at the more refined levels, this wirelength-based cost function is replaced by a congestion-based cost function that is computed as the sum of the squares of the wiring usages for all the routing bins in the design. This cost function is equivalent to the weighted sum of the wirelengths for all the nets, with the weight of a segment of a net lying in a given bin being the wiring usage of that bin.

The discontinuity in the cost function arising while moving from one level of the multilevel framework to the next, more refined level is minimized using a "density bin hierarchy", formed by recursively merging adjacent bins at a given level in order to generate the bins at the next level (thus eliminating grid boundary mismatch errors across different levels). Furthermore, a cluster is moved into a bin only if the cell area overflow in each of its ancestor bins (at coarser levels in the multilevel framework) does not exceed a given bound as a result of the move. Therefore, an overflowing bin always implies the presence of some less congested bin(s) in its neighborhood (due to the overflow constraint on the common ancestor of these bins). As a result, a coarse placement can be refined all the way to the finest level without significant area overflow.

The placement refinement scheme at each level of the refinement stage explores a large number of placement moves within its simulated annealing framework, and is therefore somewhat slow. It has been improved in [LXK+04] by the use of a greedy cell relocation scheme that considers only a small number of local moves for each cell, similar in spirit to that described in Algorithm 8. Specifically, the nets are ordered in decreasing order of the sum of their routing demands in congested cells. Then, for each net in this sorted list taken in order, cell moves are evaluated for each of the cells connected to that net. A cell move is evaluated by attempting to move the cell to every other bin within a predetermined neighborhood of the current bin of that cell, invoking the congestion-aware LZ-router to evaluate the move.

The multilevel approach described above relies on very fast incremental routing to make its fine-grained interleaving feasible from a runtime perspective. While its ability to evaluate every placement move for its congestion impact does allow for the maintenance of an accurate congestion map within a given level, the placement decisions can still be locally suboptimal because of the error arising from the coarsening and refinement process, and because of the approximations inherent in the fast interconnect topology generation. In general, although a fast router that has fidelity with the final routing is desirable within any interconnect-aware placement algorithm, the evaluation of each placement move using actual routing is usually not very cost-effective. It is often better to use fast approximations of the routing to keep an approximate congestion map, periodically restoring the accuracy of the map by invoking a fast global router.

5.3.2 Interleaved Update of Control Parameters in Congestion-aware Placement

Strictly speaking, this class of techniques consists of hybrids between the interleaved schemes discussed in Section 5.3 and the explicit congestion management schemes described in Section 5.4. Even congestion management schemes that manage congestion within the placement engine can benefit from a periodic update of the congestion model during the placement process. Since congestion is managed continuously within these schemes, their congestion modeling and update is rather simplistic, in order to keep the runtime overhead manageable. It is usually achieved through proxies such as white space allocation or net weights. Therefore, the periodic invocation of an accurate congestion evaluation engine is useful to update these proxies to bring them in line with the actual congestion that they are trying to capture. Furthermore, if this interleaved invocation of the congestion evaluation engine is not too frequent, the runtime overhead is also not excessive. Examples of this class of techniques include white space management techniques such as [HYH+01] and [BR03], and are discussed in more detail in Section 5.4.

5.4 Explicit Congestion Management within Placement

The congestion management techniques discussed within this section attempt to model and ameliorate congestion continuously during (all or part of) the global placement process. This is achieved either by presenting a slightly modified, congestion-friendly placement problem to a standard global placer, as with inflated cells or the addition of free cells, or by incorporating a fast congestion estimate into the objective function that is optimized by the placer. In order to avoid excessive overheads due to the accurate and continuous estimation of congestion in the design, these schemes model and manipulate congestion indirectly through fast proxies such as white space distribution, pin density balance constraints, or net weights. Consequently, while they are effective at improving the overall congestion profile of a design, they cannot ensure that the layout generated by them is entirely congestion-free. As a result, these techniques can benefit from the application of some post-processing scheme for congestion mitigation (as in [HYH+01]), or from the interleaved invocation of a more accurate and compute-intensive congestion estimation engine that is then used to update the congestion proxies used by these schemes (as in [HYH+01] and [BR03]).

5.4.1 Cell Inflation

Approaches based on *cell inflation*, also known as *cell bloating* or *padding*, have been used extensively for congestion relief in industrial placement engines. They rely on the observation that much of the congestion problems within a design are local and arise from difficulties in pin accessibility for the cells lying in the congested region. In other words, the congestion impact of wires having pins in a congested region is usually far more significant than that of flyover wires that are merely routed over that region. This is because the via stacks required by the local wires in order to reach their pins can create significant blockages to routing. Furthermore, unlike local wires, flyover wires can be rerouted through less congested regions. Consequently, reducing the local pin density in a congested region can be an effective way of improving its routability. Cell inflation approaches achieve this by artificially increasing the "virtual" size of the cells in congested regions, so that fewer of those cells can be placed there, resulting in a lower pin density (as depicted in Figures 5.11 and 5.12). While similar techniques have been used by designers for quite a while, they were first proposed in design automation literature in [SST+97], albeit in the context of metal programmable gate arrays. They were subsequently independently presented for standard cell based designs in [HYH+01] and [BR03].

The work in [SST+97] proposed cell inflation for congestion relief within a simulated annealing framework. A valuable insight in this work is the use of a supra-linear, monotonically increasing function of the congestion as the objective function for the placement improvement. In particular, they select $(\max\{0, c\})^2$, where c is the difference between the routing demand and supply in any given region expressed in terms of routing tracks per unit area, as the basis for their objective function. Another contribution is the derivation of a first-order expression to determine the congestion impact of a cell move involving padded cells (using a blockage-aware net bounding box model derived from [Che94]), although some of the underlying assumptions about the uniform distribution of blockages are rather simplistic. This model allows them to determine the extent of padding required for each cell. In contrast, the subsequent cell inflation works, [HYH+01] and [BR03], rely on more empirical schemes to determine the extent of inflation required for each cell.

The work in [HYH+01] embeds cell inflation into a GORDIAN-like [KSJ+91, YHQ+98] partitioning-based quadratic placement algorithm (although the center of gravity constraints described in Equation (5.3) are solved using Lagrangian relaxation rather than by reducing the rank of the system matrix by embedding them directly as in [KSJ+91]). In order to estimate the congestion, multipin nets are decomposed into two-pin Steiner segments using a star model, with each of the two single bend ("L") routes for a two-pin net or Steiner segment being considered equally likely (with some adjustment for overlapping Steiner segments belonging to the same net). The routing demand thus computed is then compared against the routing supply (after

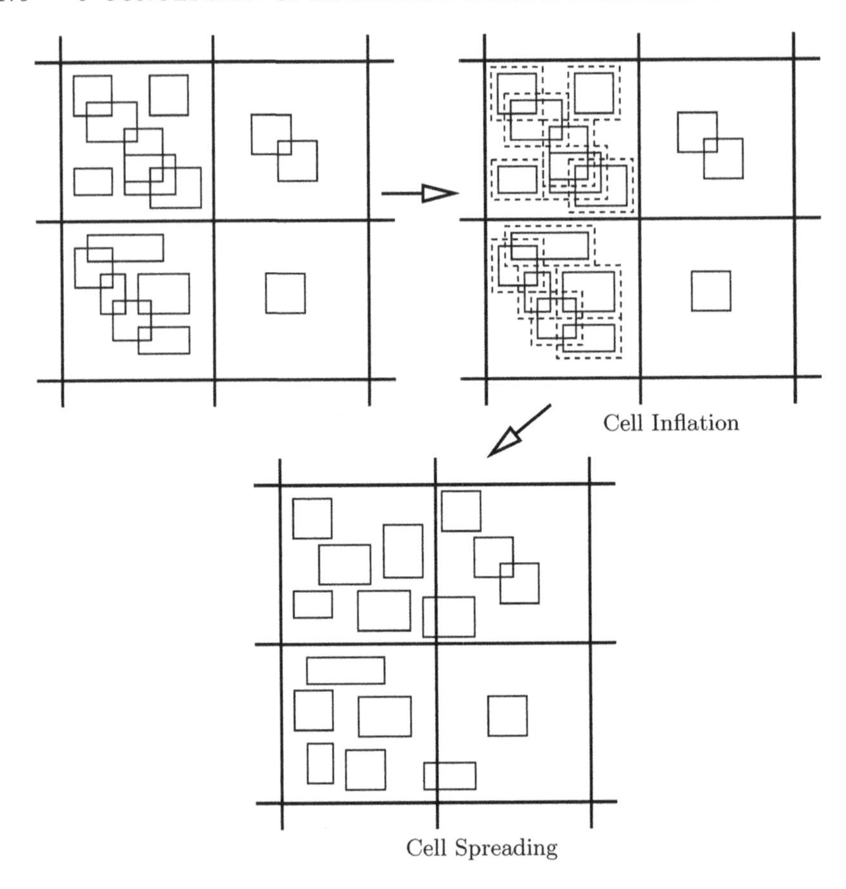

Fig. 5.11. Cell inflation in the context of global placement.

compensating for routing blockages) in order to determine the congestion in each partition. This congestion estimation is carried out for the first time at the end of the partitioning iterations during the first pass of global placement. Then, the cells in congested partitions are empirically inflated, and the quadratic placement is repeated for the previous k iterations of the partitioning using the inflated cell sizes (The implementation in [HYH+01] used $k = 6$). The redoing of a few partitioning iterations with inflated cell sizes in congested partitions tends to move cells out of these partitions and into sparsely populated partitions, even as it optimizes the global objective of the wirelength while obeying the cell spreading (*i.e.*, center of gravity) constraints. This process of cell inflation in congested partitions followed by the redoing of the previous k iterations of the partitioning of the quadratic placement is repeated a few times, until the congestion problem is solved (or a limit on the number of repetitions is reached).

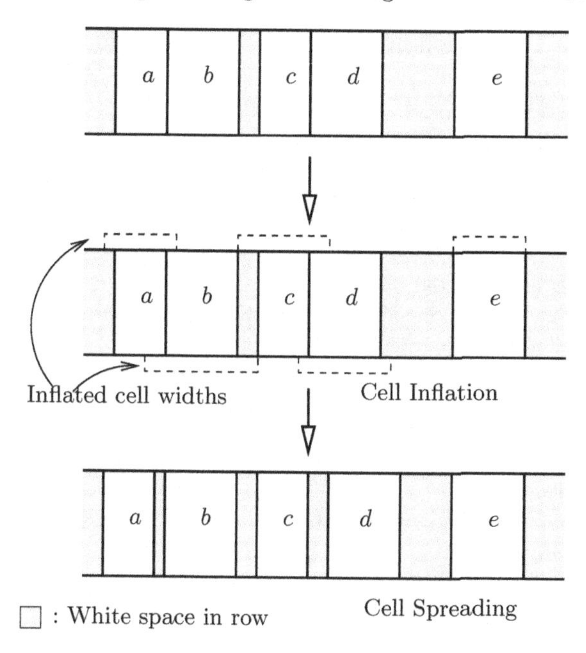

Fig. 5.12. Cell inflation within a row in the context of detailed placement.

The cell inflation phase in [HYH+01] is followed by a post-processing phase of greedy congestion optimization during which a series of ripple moves relocates cells from congested partitions to sparse ones. A straight line trajectory is drawn from the most congested global placement bin to the least congested one, and cells are moved greedily to adjacent bins along the direction of this trajectory starting from the most congested bin. This process of drawing straight line trajectories and moving cells along them is repeated a few times before the final legalization is invoked. This entire congestion-driven placement approach is summarized in Algorithm 12.

The work in [BR03] uses cell inflation in the context of the quadrisectioning-based quadratic placement algorithm BonnPlace [Vyg97]. It presents a detailed empirical scheme to determine the extent of inflation required for each cell. At any stage of the placement, the congestion due to the inter-partition nets is estimated by constructing Steiner topologies for them, and then spreading each two-pin net or Steiner segment probabilistically over all possible two-bend ("LZ") routes, while that due to the intra-partition nets is approximated by the pin density of the cells within that partition.

Let the inflated size of cell c be $(1+b(c))s(c)$, where $s(c)$ is the actual size of the cell, and $b(c) \geq 0$. For any partition P with size $s(P)$, the total inflated size of the cells within P (*i.e.*, $\sum_{c \in P}(1+b(c))s(c)$) is upper-bounded by $s(P)$. For any cell c, the initial value of $b(c)$ is proportional to the pin density of that cell (*i.e.*, the number of pins in c divided by its area $s(c)$). These values

Algorithm 12 Congestion-driven placement based on cell inflation as in [HYH+01]

1: **while** further slicing partitioning desired **do**
2: Carry out quadratic placement iteration
3: **end while**
4: Estimate congestion in each global placement bin
5: **while** congestion still present and bound on cell inflation iterations not exceeded **do**
6: Inflate cells in congested partitions
7: Redo the last k iterations of slicing partitioning for quadratic placement using inflated cell sizes
8: Estimate congestion in each global placement bin
9: **end while**
10: **if** congestion still present **then**
11: **while** congestion still present and bound on greedy post-processing iterations not exceeded **do**
12: Ripple-move cells along trajectory from most congested bin to least congested bin
13: Estimate congestion in each global placement bin
14: **end while**
15: **end if**
16: Legalize placement

are normalized so that, over the entire design, $\sum_c b(c) \cdot s(c) = \frac{\tau}{4} \sum_c s(c)$, where τ is an input parameter controlling the extent of the inflation. In other words, the total combined budget for the initial inflation of all the cells in the design is obtained using a factor of $\tau/4$. (The implementation in [BR03] uses $\tau = 0.2$). As the placement proceeds, the value of $b(c)$ for each cell c is updated depending on the local congestion of the partition containing that cell. For each of the four bounding edges of the partition containing c that is congested, $b(c)$ is increased by $\min\{1, 2(\chi(e) - 1)\} \cdot \frac{\tau}{5}$, where $\chi(e)$ is the normalized congestion of the bounding edge e, defined as the ratio of its expected demand to its capacity. This can lead to a maximum increase in $b(c)$ of $4\tau/5$ (if the congestion of each edge is at least 1.5). The remaining potential increase of $\tau/5$ in $b(c)$ is determined by the pin density of the partition. More specifically, if the pin density of the partition is greater than some threshold, the value of $b(c)$ is increased by a quantity proportional to the pin density and upper bounded by $\tau/5$. On the other hand, if each of the four bounding edges for the partition containing the cell c show no congestion problems and the pin density of the partition lies below the selected threshold, $b(c)$ is reduced by a quantity proportional to the pin density and congestion that is again upper-bounded by τ. Observe that the values $b(c)$ for all the cells within a partition increase or decrease by the same amount.

In contrast to [HYH+01] that uses inflated cell sizes by redoing the last few partitioning iterations of the quadratic placement using the inflated sizes,

Algorithm 13 Congestion-driven placement based on cell inflation within the BonnPlace framework

1: **for all** cells c in design **do**
2: Set $b(c)$ proportional to pin density of c
3: **end for**
4: **while** further quadrisectioning desired **do**
5: Carry out quadratic placement iteration
6: **for all** 2×2 windows W of adjacent partitions in design **do**
7: Quadrisection W using wirelength cost function and inflated cell sizes
8: Compute congestion on each partition boundary within W
9: Compute pin density for each partition within W
10: **for all** cells c lying inside partitions within W **do**
11: Update $b(c)$ based on partition congestion and pin density
12: **end for**
13: Redo quadrisectioning of W using wirelength cost function and updated inflated cell sizes
14: **end for**
15: Sort all 2×2 windows of adjacent partitions in design by difference between maximum and average partition congestion
16: **for all** 2×2 windows W of adjacent partitions in design, taken in sorted order, **do**
17: Locally repartition W using weighted sum of wirelength and maximum congestion within W as cost function
18: **if** local repartitioning accepted **then**
19: Update sorting key of all windows intersecting with W and reorder sorted list of windows
20: **end if**
21: **end for**
22: **end while**
23: Legalize placement

the approach in [BR03] does not redo any of the prior global placement iterations. Instead, it uses the inflated cell sizes in a local repartitioning step that moves cells out of dense partitions (as measured using the inflated cell sizes) and into adjacent partitions that are sparse. This step fits naturally into the BonnPlace framework, since that algorithm also uses a repartitioning step over all 2×2 windows in the design after each quadrisectioning iteration, for local improvements to the placement. In the congestion-aware version, this repartitioning step is also used to sort all the 2×2 windows that contain at least one congested partition, in descending order of the difference between the maximum and average congestions within the window (since a large value of this sorting key implies a locally congested partition with sparse adjacent partitions, so that the local congestion can be eased without large cell movements). The objective function for the repartitioning is a weighted sum of the wirelength and the maximum congestion within the window. Every time a repartitioning of a window is accepted, the sorting keys for all the windows containing any of the updated partitions are also updated. This entire congestion-driven placement approach is summarized in Algorithm 13.

5.4.2 White Space Management Techniques

As we have seen earlier in Section 5.2.3, the intelligent allocation of white space in a design can improve its routability significantly. The techniques in Section 5.2.3 focused on white space manipulations after the global placement had been completed. In contrast, the white space management techniques described here manage the white space continuously throughout (all or part of) the global placement process. Although these techniques are quite effective at mitigating congestion problems, they suffer from the same drawbacks as cell inflation techniques, in that they ignore non-local flyover wires and use cell density as a proxy for routability. Furthermore, they do not monitor the routing congestion continuously, and can therefore benefit from the subsequent application of some post-processing scheme for congestion mitigation. Although the cell inflation schemes described in Section 5.4.1 can also be thought of as white space management schemes operating concurrently with global placement, the techniques discussed in this section operate upon white space as an entity by itself, without associating it with individual cells. In other words, these schemes determine how much white space should be allocated to a given region in a top-down fashion, rather than building up the white space profile cell by cell in a bottom-up manner (as was the case with the techniques discussed in Section 5.4.1). Furthermore, the management of white space in these techniques is explicit, in contrast to the interleaved quadratic placement and routing scheme of [PBS98] discussed earlier in Section 5.3.1 that manipulates the white space implicitly when it adjusts the partition sizes in response to their congestion.

Some of the early white space allocation work concurrent with global placement [CKM03] was carried out in the context of the widely used academic placer CAPO [CKM00], discussed in Section 5.1.2, that is based on the recursive bisectioning paradigm using Fiduccia-Mattheyses style move heuristics for min-cut hypergraph partitioning. While routability is not the sole focus of this work, it provides a theoretical basis for the management of white space in recursive partitioning based placers. For a given region, CAPO uses horizontal or vertical cuts depending on the aspect ratio of the region (with the cut being made to partition its longest side). Vertical cuts are made with a fixed relative tolerance τ (*i.e.*, the difference between the upper and lower bounds on the cell areas in a child subregion, divided by the total cell area within the parent region), with the cut line being shifted after the partitioning (when the total cell area in each partition is available) in order to equalize the relative white space in each subregion. In contrast, since a horizontal cut line cannot be arbitrarily shifted due to the fixed height row-based structure of designs, horizontal cuts use a hierarchical white space allocation scheme that is described next.

The core concept in this work is that of *white space deterioration* during a partitioning cut. Let a region with site area S and cell area C be partitioned by a horizontal cut into two subregions with site areas S_0 and S_1 and core areas C_0 and C_1, respectively (where $S = S_0 + S_1$ and $C = C_0 + C_1$). Then, the absolute white space W of the parent region is given by $\max\{0, S - C\}$, and its relative white space w is given by W/S.The absolute and relative white spaces of the two child subregions (*i.e.*, W_0, W_1, w_0 and w_1) are defined in a similar manner. Then, this partition is said to involve a white space deterioration of α (with $0 \leq \alpha \leq 1$) if both $w_0 \geq \alpha w$ and $w_1 \geq \alpha w$. In other words, each of the subregions must have a relative white space of at least αw. The closer α is to 1, the more uniform is the distribution of white space in the design. If the permissible white space deterioration is too tight (*i.e.*, close to 1), it may not allow the partitioner to find any legal solutions, or may result in poor optimization of the wirelength because of severe restriction of the solution space (since moves of large cells may violate the local white space deterioration constraints). On the other hand, if the permissible white space deterioration is large, some partitions can end up with very little white space leading to severe local congestion problems.

The concept of white space deterioration is closely tied to the balance tolerances that the partitioner must respect. If $C_i^{\max}$ and $C_i^{\min}$ $(i = 0, 1)$ are the upper and lower bounds on the cell area C_i of the two child subregions that the partitioner must obey, it can be shown that:

$$C_i^{\max} = \min\{C, (1-\alpha)S_i + \alpha \frac{C}{S} S_i\},$$

and,

$$C_i^{\min} = \max\{0, C - C_{1-i}^{\max}\},$$

where $i = 0, 1$. Furthermore, the white space deterioration can be expressed in closed form in terms of the relative white space w in a block with R rows as:

$$\alpha = \frac{\sqrt[n+1]{1-w} - (1-w)}{w \sqrt[n+1]{1-w}},$$

where $n = \lceil \lg_2 R \rceil$. Thus, the permissible white space deterioration increases as the partitioner descends to lower levels (*i.e.*, the number of rows in the block being partitioned decreases). Coupled with the cut line shifting for vertical cuts, this approach finds improved cut sizes while facilitating good use of white space when it is scarce, and spreading it uniformly when it is abundant, as compared to the case when the partitioning tolerance is constant across all levels. As a corollary, this usually leads to routability improvements also.

We have seen in Section 5.2.3 that the intelligent manipulation of white space can improve the routability of a design considerably. For placers such as CAPO that tend to distribute white space uniformly, some of these routability improvements can be achieved by the introduction of a small number of *free cells*, as proposed in [AML06]. Free cells are small, disconnected cells added to a design before placement and removed from it before the routing. They allow the placer to improve the wirelength of the placement by increasing the "virtual tolerance" in the deviations from uniform white space allocation. Free cells are even more effective in improving placement quality without impacting routability when used with a congestion-driven placer such as DRAGON [YCS03]. However, if the free cells use up too large a fraction of the available white space, they can constrain the placer excessively and lead to unroutable layouts.

In general, white space allocation techniques used concurrently with the global placement can improve the routability of the design. However, because of the inaccurate congestion estimation available early in the global placement phase, the routability impact of techniques that choose the control parameters for their white space allocation strategy very early during placement is limited. A natural research avenue would be to investigate concurrent global placement and white space allocation techniques that dynamically adapt their white space allocation strategy as the global placement proceeds. Such techniques are expected to combine the large search space available to the white space allocation techniques discussed above with the effectiveness of the techniques discussed in Section 5.2.3.

5.4.3 Congestion-aware Objective Function or Concurrent Constraints

Several techniques try to model some fast, dynamically updated proxy for the congestion as part of the objective function that the global placement process seeks to optimize, or as part of the set of constraints that the placer must obey. Indeed, some of these techniques form a continuum of hybrids with interleaved schemes (as discussed in Section 5.3) that update the dynamically modeled proxies periodically with some more accurate and compute-intensive congestion estimation engine. Thus, for instance, the interleaved quadratic placement and routing scheme of [PBS98] models congestion into the net weights (and thus, indirectly into the objective function which is the weighted sum of the quadratic netlengths)[3]. In contrast, the multilevel placement scheme of [CCP+03] models congestion as part of the objective function for simulated annealing when operating in the congestion-driven mode (*i.e.*, at the finer levels of refinement).

There were several early efforts to model congestion continuously into the placement process. For instance, [ML90] used a lookup table of precomputed Steiner trees to quickly determine the congestion cost of each cell move within a partitioning-based placer, thus constraining the set of feasible cell moves. Within a preprocessing step, one or more "good" Steiner trees were computed

[3] The modeling of the congestion using net weights is also used within the SPARSE [HM02] scheme discussed in Section 5.2.2, although this is done as a post-processing step after global placement.

for every possible distribution of the pins of a net across the partitions of some given level of partitioning. However, this approach is not scalable because of two reasons. Firstly, precomputed Steiner trees cannot be modified dynamically in response to varying congestion profiles or performance constraints. Secondly, because of its exhaustive nature, the size of the lookup table quickly becomes prohibitive (the work in [ML90] restricts the total number of partitions in the design to sixteen – a granularity that is excessively coarse for today's typical flat placement problems that often involve more than 10^6 cells). Another early yet influential work [Che94] introduced the RISA blockage-aware net bounding box model[4] for quick congestion computation in order to embed it into the objective function of a simulated annealing based placer.

Congestion Control using a Temporary Constraint Relaxation Framework

Another approach to handling congestion was presented in [ZD02] as part of the SPADE [ZD00] partitioning-based placement framework for systematic constraint optimization. In this work, different placement objectives are modeled as different classes of constraints, whose temporary relaxation is permitted in the intermediate global placement stages, as long as these violations can be eliminated in the final placement solution. This intermediate relaxation of the constraints is very effective at helping the placer escape from local minima in the search space.

Constraints are classified into *balance* constraints and *non-balance* constraints, where balance constraints require the ratios of the metric values modeled by the constraints for the two subpartitions created by the partitioning cut to lie within a small range. Thus, for instance, a balance constraint on the sum of the cell sizes within the two partitions could require that these sums be approximately equal (with the approximate equality being specified formally as a set of tolerances). Formally, if the metric values for the constraint of interest are V, V_0, and V_1 for the original partition and the two subpartitions, respectively, a balance constraint could require that:

$$V_0/V \leq r + t_0,$$

and,

$$V_1/V \leq 1 - r + t_1,$$

where $r : (1 - r)$ is the specified ratio between the values of the metric across the two partitions, and t_0 and t_1 are the permissible tolerances for these values with which the constraint must be satisfied.

[4] An important contribution of this work, discussed in more detail in Section 2.1.1 in Chapter 2, was the empirical evaluation of the factor by which the actual wirelength of a multipin net could be expected to exceed its simple half-rectangle perimeter bounding box prediction. This factor grew slowly from 1.0 for two or three pin nets to 2.7933 for fifty pin nets.

Different classes of constraints are combined together using *normalized cell constraint weights* or *c-weights*. Each cell has a vector of c-weights whose i^{th} element represents the change in the normalized metric values of the ith constraint as a result of a move of that cell across the current partitioning cut. A temporary constraint relaxation is permitted only if (i) a legal solution can be reached within a given number of steps even after the proposed violation, and (ii) the best cost solution achievable without the violation is more expensive than the best cost legal solution reachable after the relaxation. Furthermore, in the context of multiple constraints, a new constraint violation is permitted only if it does not worsen the currently relaxed constraint that is deemed most critical (on the basis of the c-weights, although comparing the normalized impact of different classes of constraints can be difficult). The work in [ZD02] presents a mathematical framework for checking these conditions in order to decide when a constraint relaxation can be permitted.

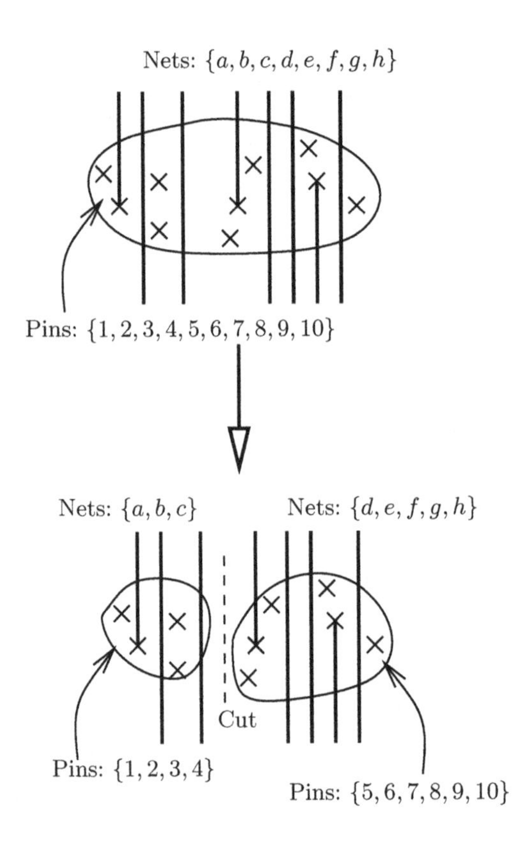

Fig. 5.13. Modeling congestion through balance constraints on pin density and external net density.

Congestion is modeled using two sets of balance constraints across each partitioning cut, namely, pin density constraints and parallel external net constraints, as depicted in Fig. 5.13. The external nets are nets that do not cross the cut, and include the flyover nets over the region. Their recursive allocation to one of the two newly created partitions at each partitioning cut is equivalent to global route planning. Handling the external net balance constraint requires keeping track of the crossing points of all the nets across the cuts, in a manner somewhat similar to the hierarchical net topology generation approach proposed in [SK89] (except that situations in which no single net topology is clearly the best do not require premature commitment to a single topology; instead, choice is preserved by assigning probabilities to each of the topology options, with the congestion being computed by weighting each of these options with their probabilities). The recursive balancing of the structural pin densities and the external net densities across all the partitions helps avoid local congestion hot spots.

5.5 Final Remarks

In this chapter, we have looked at a wide class of techniques for congestion relief during placement. Placement is indeed the workhorse of congestion mitigation in today's industrial physical synthesis flows. While congestion is almost never the primary metric driving placement, the placement phase still remains the most effective stage during design implementation for improving routability. Most of the congestion relief techniques in commercial use today fall into the classes of post-processing techniques or interleaved techniques, although cell inflation has also proven quite effective at relieving pin accessibility problems in local congestion hot spots.

Many of the techniques described in this chapter were presented as case studies in the contexts of the placement algorithms within which they have been implemented. However, these techniques are often independent of their specific implementation and can be easily adapted to other placement paradigms also. Thus, for instance, the row-based white space allocation used in DRAGON is equally applicable to global placements produced by analytical placers, while the region size adjustment scheme based on interleaved routing and quadratic placement described in [PBS98] can be used for partition size adjustment in partitioning-based placers also.

While much work has already been done in post-processing and interleaved congestion mitigation schemes, there is still need for further research in the area of congestion relief that operates concurrently with global placement using fast, accurate congestion proxies, periodically updated in an interleaved fashion, that are part of the objective function. A promising direction in this context is the recent work on white space allocation; further exploration of this area is likely to yield promising results on achieving enhanced routability without degrading the primary placement metrics significantly.

References

[AML06] Adya, S. N., Markov, I. L., and Villarrubia, P. G., On whitespace and stability in physical synthesis, *Integration — The VLSI Journal* 39(4), pp. 340–362, 2006.

[BR03] Brenner, U., and Rohe, A., An effective congestion-driven placement framework, *IEEE Transactions on Computer-Aided Design of Integrated Circuits and Systems* 22(4), pp. 387–394, April 2003.

[Bre77] Breuer, M., Min-cut placement, *Journal of Design Automation of Fault Tolerant Computers* 1(4), pp. 343–362, Oct. 1977.

[CKM00] Caldwell, A. E., Kahng, A. B., and Markov, I. L., Can recursive bisection alone produce routable placements?, *Proceedings of the Design Automation Conference*, pp. 477–482, 2000.

[CKM03] Caldwell, A. E., Kahng, A. B., and Markov, I. L., Hierarchical whitespace allocation in top-down placement, *IEEE Transactions on Computer-Aided Design of Integrated Circuits and Systems* 22(11), pp. 1550–1556, Nov. 2003.

[CCK+00] Chan, T., Cong, J., Kong, T., and Shinnerl, J., Multilevel optimization for large-scale circuit placement, *Proceedings of the International Conference on Computer-Aided Design*, pp. 171–176, 2000.

[CCP+03] Chang, C.-C., Cong, J., Pan, Z., and Yuan, X., Multilevel global placement with congestion control, *IEEE Transactions on Computer-Aided Design of Integrated Circuits and Systems* 22(4), pp. 395–409, April 2003.

[Che94] Cheng, C.-L. E., RISA: Accurate and efficient placement routability modeling, *Proceedings of the International Conference on Computer-Aided Design*, pp. 690–695, 1994.

[CK84] Cheng, C., and Kuh, E., Module placement based on resistive network optimization, *IEEE Transactions on Computer-Aided Design of Integrated Circuits and Systems* CAD-3(3), pp. 218–225, July 1984.

[CV04] Chu, C., and Vishwanathan, N., FastPlace: Efficient analytical placement using cell shifting, iterative local refinement, and a hybrid net model, *Proceedings of the International Symposium on Physical Design*, pp. 26–33, 2004.

[CSX+05] Cong, J., Shinnerl, J. R., Xie, M., Kong, T., and Yuan, X., Large-scale circuit placement, *ACM Transactions on Design Automation of Electronic Systems* 10(2), pp. 1–42, April 2005.

[DK85] Dunlop, A., and Kernighan, B., A procedure for placement of standard-cell VLSI circuits, *IEEE Transactions on Computer-Aided Design of Integrated Circuits and Systems* CAD-4(1), pp. 92–98, Jan. 1985.

[EJ98] Eisenmann, H., and Johannes, F. M., Generic global placement and floorplanning, *Proceedings of the Design Automation Conference*, pp. 269–274, 1998.

[FM82] Fiduccia, C. M., and Mattheyses, R. M., A linear-time heuristic for improving network partitions, *Proceedings of the Design Automation Conference*, pp. 175–181, 1982.

[HYH+01] Hou, W., Yu, H., Hong, X., Cai, Y., Wu, W., Gu, J., and Kao, W. H., A new congestion-driven placement algorithm based on cell inflation, *Proceedings of the Asia and South-Pacific Design Automation Conference*, pp. 605–608, 2001.

[HM02] Hu, B., and Marek-Sadowska, M., Congestion minimization during placement without estimation, *Proceedings of the International Conference on Computer-Aided Design*, pp. 739–745, 2002.

[HL00] Hur, S.-W., and Lillis, J., Mongrel: Hybrid techniques for standard cell placement, *Proceedings of the International Conference on Computer-Aided Design*, pp. 165–170, 2000.

[JL04] Jariwala, D., and Lillis, J., On interactions between routing and detailed placement, *Proceedings of the International Conference on Computer-Aided Design*, pp. 387–393, 2004.

[KRV02] Kahng, A. B., Rodman, P., and Villarrubia, P. G., Physical chip implementation methodology: Hot spots and best practices, *Tutorial at Design Automation Conference*, 2002, available at `http://vlsicad.ucsd.edu/~abk/TALKS/`.

[KW04] Kahng, A. B., and Wang, Q., Implementation and extensibility of an analytical placer, *Proceedings of the International Symposium on Physical Design*, pp. 18–25, 2004.

[Kar99] Karypis, G., Multilevel algorithms for multi-constraint hypergraph partitioning, *Technical Report* 99-034, Department of Computer Science, University of Minnesota, Minneapolis, MN, 1999.

[KSJ+91] Kleinhans, J. M., Sigl, G., Johannes, F. M., and Antreich, K., GORDIAN: VLSI placement by quadratic programming and slicing optimization, *IEEE Transactions on Computer-Aided Design of Integrated Circuits and Systems* 10(3), pp. 356–365, March 1991.

[LXK+04] Li, C., Xie, M., Koh, C.-K., Cong, J., and Madden, P. H., Routability-driven placement and white space allocation, *Proceedings of the International Conference on Computer-Aided Design*, pp. 394–401, 2004.

[ML90] Mayrhofer, S., and Lauther, U., Congestion-driven placement using a new multi-partitioning heuristic, *Proceedings of the International Conference on Computer-Aided Design*, pp. 332–335, 1990.

[NDS01] Naylor, W., Donelly, R., and Sha, L., Non-linear optimization system and method for wire length and delay optimization for an automatic electric circuit placer. *U.S. Patent* 6301693, Oct. 2001.

[PBS98] Parakh, P. N., Brown, R. B., and Sakallah, K. A., Congestion driven quadratic placement, *Proceedings of the Design Automation Conference*, pp. 275–278, 1998.

[SST+97] Sadakane, T., Shirota, H., Takahashi, K., Terai, M., and Okazaki, K., A congestion-driven placement improvement algorithm for large scale sea-of-gates arrays, *Proceedings of the Custom Integrated Circuits Conference*, pp. 573–576, 1997.

[SR99] Sankar, Y., and Rose, J., Trading quality for compile time: Ultra-fast placement for FPGAs, *Proceedings of the International Symposium on Field Programmable Gate Arrays*, pp. 157–166, 1999.

[SS86] Sechen, C., and Sangiovanni-Vincentelli, A., TimberWolf 3.2: A new standard cell placement and global routing package, *Proceedings of the Design Automation Conference*, pp. 432–439, 1986.

[SDJ91] Sigl, G., Doll, K., and Johannes, F., Analytical placement: A linear or quadratic objective function?, *Proceedings of the Design Automation Conference*, pp. 427–432, 1991.

[SK89] Suaris, P. R., and Kedem, G., A quadrisection-based combined place and route scheme for standard cells, *IEEE Transactions on Computer-Aided Design of Integrated Circuits and Systems* 8(3), pp. 234–244, March 1989.

[TCT92] Tsay, R.-S., Chang, S. C., and Thoraldson, J., Early wirability checking and 2-D congestion-driven circuit placement, *Proceedings of the International Conference on ASIC*, pp. 50–53, 1992.

[THK88] Tsay, R.-S., Kuh, E., and Hsu, C., Proud: A fast sea-of-gates placement algorithm, *IEEE Design and Test of Computers*, pp. 44–56, 1988.

[Vyg97] Vygen, J., Algorithms for large-scale flat placement, *Proceedings of the Design Automation Conference*, pp. 746–751, 1997.

[WYE+00] Wang, M., Yang, X., Eguro, K., and Sarrafzadeh, M., Multi-center congestion estimation and minimization during placement, *Proceedings of the International Symposium on Physical Design*, pp. 147–152, 2000.

[WYS00a] Wang, M., Yang, X., and Sarrafzadeh, M., Congestion minimization during placement, *IEEE Transactions on Computer-Aided Design of Integrated Circuits and Systems* 19(10), pp. 1140–1148, Oct. 2000.

[WYS00b] Wang, M., Yang, X., and Sarrafzadeh, M., Dragon2000: Fast standard-cell placement for large circuits, *Proceedings of the International Conference on Computer-Aided Design*, pp. 260–263, 2000.

[YCS03] Yang, X., Choi, B.-K., and Sarrafzadeh, M., Routability-driven white space allocation for fixed-die standard-cell placement, *IEEE Transactions on Computer-Aided Design of Integrated Circuits and Systems* 22(4), pp. 410–419, April 2003.

[YWK+03] Yang., X., Wang, M., Kastner, R., Ghiasi, S., and Sarrafzadeh, M., Congestion reduction during placement with provably good approximation bound, *ACM Transactions on Design Automation of Electronic Systems*, pp 1–17, Feb. 2003.

[YM01] Yildiz, M. C., and Madden, P., Global objectives for standard cell placement, *Proceedings of the Great Lakes Symposium on VLSI*, pp. 68–72, 2001.

[YHQ+98] Yu, H., Hong, X., Qiao, C., and Cai, Y., CASH: A novel quadratic placement algorithm for very large standard cell layout design based on clustering, *Proceedings of the International Conference on Solid-state and Integrated Circuit Technology*, pp. 496–501, 1998.

[ZD00] Zhong, K., and Dutt, S., Effective partition-driven placement with simulataneous level processing and global net views, *Proceedings of the International Conference on Computer-Aided Design*, pp. 254–259, 2000.

[ZD02] Zhong, K., and Dutt, S., Algorithms for simultaneous satisfaction of multiple constraints and objective optimization in a placement flow with application to congestion control, *Proceedings of the Design Automation Conference*, pp. 854–859, 2002.

6

CONGESTION OPTIMIZATION DURING TECHNOLOGY MAPPING AND LOGIC SYNTHESIS

In Chapters 4 and 5, we have observed how the many theoretical and practical advances in congestion alleviation during the layout stages form the basis for congestion-awareness in modern physical synthesis flows. However, the congestion problem in many of today's design blocks is severe enough that these congestion alleviation techniques do not suffice, resulting in several congested blocks requiring multiple synthesis and layout iterations before their routability problems can be resolved. Often, the problem can be traced back to the synthesized netlists being suboptimal from the point of view of congestion.

As the example discussed in Section 3.1 of Chapter 3 demonstrated, synthesis choices that are area- or delay-optimal may not be optimal from the point of view of congestion, and may even lead to unroutable circuits. The logic synthesis[1] and technology mapping stages offer degrees of freedom that are not available during layout. More specifically, unlike the layout stages where the set of wires in the netlist is fixed, these stages can absorb wires within logic gates or split logic functions into smaller gates, and can therefore potentially have a large impact on the post-routing congestion. Consequently, it is useful to perform congestion optimization during these stages also.

However, it is more difficult to predict the congestion impact of an optimization choice during these stages than during placement, as was discussed in Chapter 3. Although the metrics used for congestion estimation during technology mapping and logic synthesis are in general not as accurate as the ones used during placement, there has still been much promising work in the development of netlist transformations that can provide downstream routability benefits using these metrics. This chapter reviews several such optimization techniques.

A typical synthesis flow begins with a register transfer level (RTL) description of the logic that is to be implemented, which is then translated

[1] As mentioned in Chapter 3, we will continue to use the term "logic synthesis" for the technology-independent logic synthesis stage, and refer to technology mapping explicitly when needed.

into a Boolean network. Various sequential optimizations, such as retiming and state encoding, are performed on the Boolean network. To the best of our knowledge, no results published to date have targeted congestion during these sequential optimizations, primarily because it is difficult to estimate routing congestion at such a high level of design abstraction. The sequential transformations are followed by a set of combinational optimizations; these include the application of multilevel logic synthesis operations such as substitution and the extraction of common subexpressions. Recent research has focused on making these transformations "congestion-aware" by modifying the underlying cost functions to incorporate some appropriate congestion metric. These transformations are followed by the technology decomposition step which converts a given Boolean network into the subject graph, which is a network of primitive gates such as two-input NANDs and inverters. Traditionally, this decomposition has aimed at optimizing the delay or the area, and we will discuss its extension to consider congestion as well, since the quality of the subject graph affects that of the mapping solution. Finally[2], the subject graph is implemented as a netlist that uses only the cells available in some given library, during the technology mapping stage.

Although optimizations at the more abstract levels in the design flow can have greater impact than those applied later in the flow, the fidelity and accuracy of the metrics driving these optimizations decreases as the abstraction level of the design increases. This makes it harder to control optimizations applied early in the design flow to achieve the desired end result. Although technology mapping occurs after the logic synthesis stage in the design flow, we will first focus on congestion-aware technology mapping, in Section 6.2, and then study congestion alleviation during decomposition followed by the enhancement of other logic synthesis techniques to improve routability, in Section 6.4. In this way, we will continue our trend of discussing the congestion optimization techniques applicable at a given design stage in order of increasing design abstraction. This ordering is motivated by the sequence in which these techniques have been adopted in practice. Furthermore, designers prefer to fix congestion problems at the lowest level of abstraction possible, in order to minimize the design perturbation.

6.1 Overview of Classical Technology Mapping

The technology mapping problem aimed at minimizing area is known to be NP-hard for general graphs, but can be solved optimally in polynomial time for special structures such as trees [Keu87, DeM94]. Most technology mapping algorithms such as [Keu87, CP95, SSL+92, KBS98, SIS99] use dynamic

[2] An exception to the usual approach of having separate technology decomposition and mapping steps is the work in [LWG+97], which explores the space of the algebraic decompositions and technology mapping simultaneously.

programming to optimize some given cost function, which has traditionally been area, delay, power, or some linear combination of these metrics. These algorithms proceed in two phases: *matching* and *covering*. During the matching process, non-inferior choices for the mapped cells are stored at all the nodes in a topological traversal of the circuit, whereas the covering phase proceeds backwards in a reverse topological manner starting from the primary outputs and selecting the best choice among the matches stored at each node.

The matching process can be explained using the example shown in Fig. 6.1. The network shown in the figure represents an implementation of the Boolean function $\overline{defg}$. There are five nodes in the network, namely, N_1, N_2, N_3, N_4, and N_5. Each node represents a Boolean function computed on its inputs. For instance, N_2 denotes $\overline{fg}$, whereas N_4 represents $\overline{N_2}$. Let us assume that the library contains only an inverter, a two-input NAND cell, and a four-input NAND cell. The figure shows all the matches for every node in the network. Thus, there is only one match at node N_4, which is that of an inverter[3], whereas there are two matches at nodes N_5 (namely, a two-input NAND cell and a four-input NAND cell). The matches at a given node are obtained by performing pattern matching between the subgraph rooted at the node and the patterns corresponding to cells in the library. The pattern matching can be performed either by graph isomorphism or using binary decision diagrams (BDDs) [DeM94].

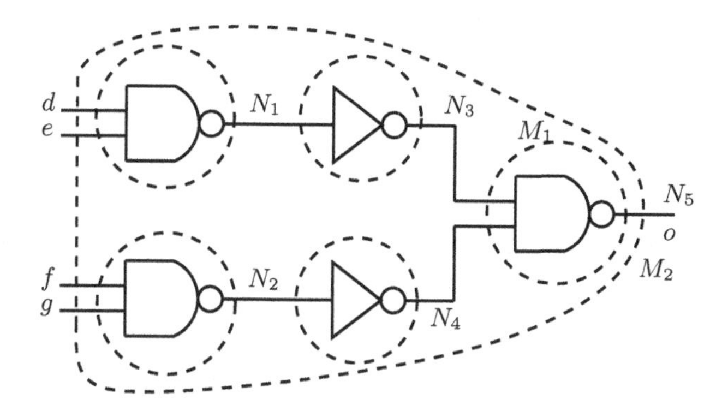

Fig. 6.1. The matching phase in technology mapping.

6.1.1 Mapping for Area

For the optimization of a relatively simple cost function such as the cell area, it suffices to store only the minimum area match at each node. The cell area

[3] We ignore the trivial match of a NAND cell with all its inputs tied together, since such a match is usually inferior to the inverter.

$A(M_{i,N_j})$ of the mapping solution due to a match M_i at a node N_j is simply the sum of the area $A(M_i)$ of the match (which is available from the library) and the sum of the areas of the optimal mapping solutions at the fanins of the match, and can be expressed as:

$$A(M_{i,N_j}) = A(M_i) + \sum_{f \in \text{fanin}(M_i)} A(f). \qquad (6.1)$$

The optimal mapping solution at a node N_j (with area denoted by, say, $A(N_j)$) is the minimum area solution among all the matches at the node, and is given by:

$$A(N_j) = \min_{M_i:\text{match at } N_j} \{A(M_{i,N_j})\}.$$

Since the matching proceeds in topological order, note that the areas of the optimal mapping solutions at the fanins of the potential matches at a node are already known when these matches are being evaluated. For instance, the area of the optimal mapping solution at N_1 in the example depicted in Fig. 6.1 is available before the mapping solutions at N_3 are evaluated. For the sake of illustration, let us assume that the areas of the inverter, the two-input NAND cell, and the four-input NAND cell are one, two, and four units, respectively. Then, the area $A(N_1)$ of the mapping solution at N_1, obtained using the match of a two-input NAND cell, is two units. As a result, the area of the best solution at N_3 is $A(N_1) + A(M_{inv}) = 2+1$, *i.e.*, three units (where M_{inv} is the inverter matching at N_3). There are two matches at N_5, namely, M_1 and M_2, resulting in an area of eight and four units, respectively. Therefore, only match M_2 is stored as the area-optimal match at node N_5.

6.1.2 Mapping for Delay

In contrast to simple cost functions like area, technology mapping for the optimization of more complicated cost functions such as delay may require multiple matches at each node during the matching phase. Delay-driven technology mapping typically uses either *load-dependent* or *gain-based* delay models. In the case of the load-dependent delay model, the area of a library cell is considered fixed, whereas its delay varies with the load that it is driving[4]. In contrast, under the gain-based delay model, the delay through the cell is considered constant across a specified range of loads, as it is assumed that the cell will be sized to meet this constant delay.

---

[4] Although the delay through a cell depends on the slews of the inputs to the cell as well as on the load being driven by the cell, the delay models used at the technology mapping level often ignore the input slews for the sake of efficiency.

Load-dependent Delay Models

Figure 6.2(a) shows typical load-delay curves specifying the load-dependent delay models for the inverter, two-input NAND, and four-input NAND cells used for the example in Fig. 6.1. The intercept on the delay axis shows the intrinsic delay of the corresponding cell. Thus, for example, the intrinsic delay of the inverter is one unit, whereas that of the two-input NAND cell is two units. The slope of the curve is proportional to the effective driving resistance of the cell; in our example, this slope is one, two, and four units for the inverter, the two-input NAND cell, and the four-input NAND cell, respectively.

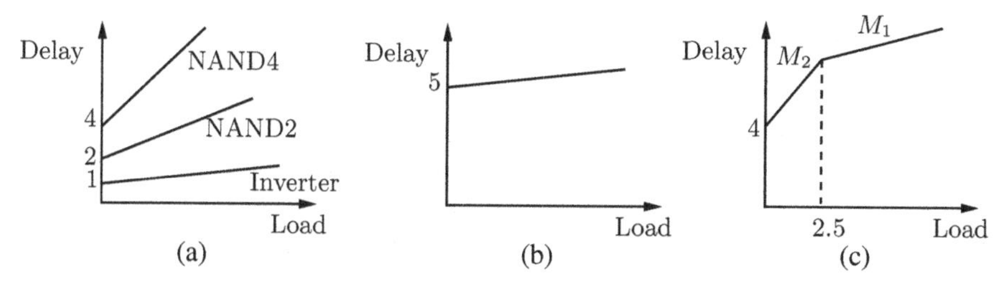

Fig. 6.2. Computation of load-delay curves: (a) Typical load-delay curves for the inverter, NAND2, and NAND4 cells. (b) The load-delay curve denoting the mapping solution due to the match at node N_3. (c) The load-delay curve representing solutions due to non-inferior delay matches for solutions at N_5.

The delay $D(M_{i,N_j}, C_L)$ of the mapping solution due to a match M_i at a node N_j when driving a load C_L is simply the sum of the delay $D(M_i, C_L)$ of the match when driving C_L and the maximum of the delays of the optimal mapping solutions at the fanins of the match when driving the match, and can be expressed as:

$$D(M_{i,N_j}, C_L) = D(M_i, C_L) + \max_{f \in \text{fanin}(M_i)} \{D(f, C_g(M_i))\}, \tag{6.2}$$

where $C_g(M_i)$ is the gate capacitance of the cell corresponding to the match M_i. The best mapping solution (with delay $D(N_j, C_L)$) at a node N_j when driving a load C_L is merely the best choice among all the mapping solutions due to the matches available at that node. More precisely,

$$D(N_j, C_L) = \min_{M_i:\text{match at } N_j} \{D(M_{i,N_j}, C_L)\}.$$

In general, the load driven by a node consists of the gate capacitance of the receiver(s) being driven by the node, and the capacitance of the wiring required to connect to the receiver(s). Although this load is unknown during the matching phase, it is available during the covering phase (since the matches are selected in the reverse topological order while being covered). At that time, it can be computed using either a wire-load model or based on the locations of the receiver(s). Therefore, the match that yields the minimum delay can be chosen during the covering phase.

The computation of the load-delay curves for the mapping solutions at N_3 and N_5 is shown in Fig. 6.2(b) and (c), respectively. For the sake of simplicity, we assume that the gate capacitances C_g of the three cells in the library are proportional to their areas, and that the capacitances of the wires connecting them are insignificant. There is only one match (namely, an inverter) at node N_3. So, the best delay $D(N_3, C_L)$ at this node (which is due to the unique mapping solution whose delay is denoted by $D(M_{inv,N_3}, C_L)$) when driving a load C_L is given by:

$$\begin{aligned} D(N_3, C_L) &= D(M_{inv,N_3}, C_L) \\ &= D(M_{inv}, C_L) + D(M_{NAND2,N_1}, C_g(M_{inv})) \\ &= (1 + 1 \times C_L) + (2 + 2 \times 1) \\ &= 5 + C_L, \end{aligned}$$

as shown in the figure. Similarly, the delay $D(N_5, C_L)$ of the best mapping solutions at node N_5 when driving a load C_L is given by the piecewise linear curve obtained by combining the delay curves for the mapping solutions due to the matches M_1 and M_2, and is given by the following equation:

$$D(N_5, C_L) = \begin{cases} 4 + 4 \times C_L, & 0 \leq C_L \leq 2.5, \\ 9 + 2 \times C_L, & 2.5 < C_L < \infty. \end{cases}$$

The above equation indicates that the mapping solution due to M_2 using the four-input NAND cell is optimal for loads of up to 2.5 units, whereas the one due to M_1 (that uses the two-input NAND cell instead) is optimal for loads greater than 2.5 units.

An implementation of a technology mapper based on the piecewise linear curve propagation discussed above is part of the publicly available sequential circuit synthesis package SIS [SSL+92].

Gain-based Delay Models

Using the gain-based delay model, the delay D of a cell is given by the following equation [SSH99]:

$$D = g \times h + p, \tag{6.3}$$

where g is the so-called *logical gain* (also referred to merely as the *gain*) of the cell, h is the so-called *electrical effort*, and p is the parasitic delay due to capacitances internal to the cell. The gain of a cell is determined by the topology of the transistor network within the cell. For a given cell, g and p are constants, whereas h is given by:

$$h = C_L/C_g, \tag{6.4}$$

where C_L is the capacitive load driven by the cell and C_g is the gate capacitance of the cell itself. If we fix h, then it implies that the cell will be sized to the value $C_g = C_L/h$ such that the delay through the cell will theoretically remain constant for all loads. The covering phase assigns a size to a matched cell using Equation (6.4) when the load driven by the cell is known. The use of gain-based delay models improves the runtime and memory complexity of technology mapping with large libraries that include multiple sizes for each cell (since multiple solutions corresponding to different sizes of the same cell do not need to be stored separately).

In practice, each cell type still has some maximum size in the library, which limits the load that can be driven by the cell. Therefore, the matching phase of a technology mapping algorithm that uses gain-based delay models may store a "delay vs. maximum load" curve [HWM03] at each node. These delay models are also better suited for delay-driven DAG mapping than load-dependent models, as discussed in the next section. More details on technology mapping with gain-based delay models can be found in [SIS99,HWM03,KS04].

6.1.3 Tree and DAG Mapping

In general, a Boolean network or subject graph is a directed acyclic graph (DAG). In this DAG, the primary inputs (outputs) have only outgoing (incoming) edges, whereas intermediate nodes are allowed to have both types of edges (representing the fanins and fanouts from the nodes). Technology mapping algorithms that operate on such a graph can be classified into *tree mapping* and *DAG mapping* algorithms, depending on whether they permit a match to subsume a multifanout point of the graph.

Tree mapping prohibits mapping across these multifanout points, and proceeds by first decomposing the DAG into trees. Each multifanout point is set to be the root of a tree, and matching and covering are performed separately on each of the trees. These decomposed trees are also called *fanout-free regions*, since all non-root nodes in the trees have a fanout of one. The creation of buffer trees may be required at the multifanout points in order to meet the load constraints at their drivers; if so, this fanout optimization is usually carried out during the covering phase or after the technology mapping.

Either of the load-dependent or the gain-based delay models can be used with tree mapping for delay optimization. However, for area optimization under delay constraints, it is usually more convenient to use the load-dependent delay model. In this case, the matching phase stores the delay-optimal choices using a piecewise linear curve and also associates the area with each of the stored solutions. Then, during the covering, the minimum area choices from among those that satisfy the delay constraints can be chosen at the primary outputs and propagated backwards. In contrast, area optimization under delay constraints is relatively difficult when used with gain-based delay models, although heuristics such as the global gain can be used for this purpose [HWM03].

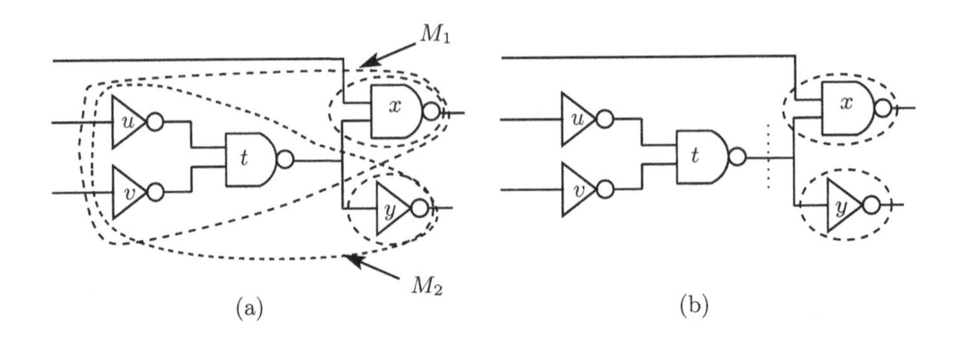

Fig. 6.3. (a) Matches across multifanout points in DAG mapping that can potentially result in logic replication. (b) Tree mapping does not allow any matches across multifanout points.

One of the disadvantages of tree mapping is that it introduces suboptimality by prohibiting choices across the multifanout points. DAG mapping overcomes this limitation by allowing matches to cross the multifanout points. Figure 6.3(a) shows an example of DAG mapping, where matches such as M_1 and M_2 across the multifanout point driven by the two-input NAND gate t are permitted. Such choices are stored during the matching phase and are selected during the covering phase if they lead to optimal delays. This, however, may involve an area penalty due to logic replication. For example, if both of the matches M_1 and M_2 are selected during the covering phase, then the functionality of the gates u, v, and t is replicated in the two matches. In the case of tree mapping, shown in Fig. 6.3(b), such matches are prohibited; thus, there is only one possible match (namely, the two-input NAND gate) at node x, since the matches are not allowed to cross multifanout points. This can lead to a suboptimal delay solution, since the search space explored is smaller when compared to that explored during DAG mapping.

The logic replication that can occur during DAG mapping creates another complication if the mapping process relies on load-dependent delay models. A single fanout node may become a multifanout one due to its receiver being replicated. This creates a problem in the computation of the load-delay curves during the matching phase. For example, let us assume that nodes driving the inputs to the gates u and v have no other fanouts, and that the load-delay curves for these nodes have already been constructed. Now, in order to compute the load-delay curve for the match M_1 at x, we need to know the load driven by its fanins, as is evident from Equation (6.2). However, this load may be unknown, since in DAG mapping, this load depends on the match M_2 at y also, which may not yet have been processed. To overcome this problem, it is usually more convenient to use gain-based delay models with DAG mapping approaches [KBS98,HWM03,SIS99], since these models assume that the delay of a match is constant, independent of its load.

6.2 Congestion-aware Technology Mapping

The classical technology mapping algorithms discussed in the previous section have been extended to incorporate congestion awareness in several ways. These extensions rely on placement-dependent congestion metrics such as the netlength [PPS03, KS01] and congestion maps [SSS+05, SSS06], or on graph theoretic ones such as the mutual contraction [LM05]; these metrics have been discussed in detail in Chapter 3. Although the mapping algorithm that uses mutual contraction does not require the prior placement of the subject graph, those that rely on placement-dependent metrics derive the location of each match in the circuit graph from the subject graph placement.

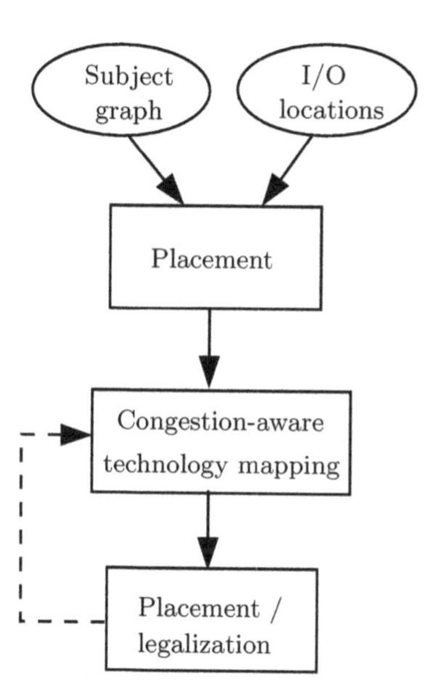

Fig. 6.4. A congestion-aware technology mapping flow that uses the placement of the subject graph to estimate the congestion during the technology mapping.

A typical design flow for an algorithm that uses the subject graph placement to estimate the congestion is shown in Fig. 6.4. The dashed feedback arrow represents the possibility of generating multiple mapped versions of the netlist (by using different variants of the cost function), and then selecting the best version for subsequent layout. The flow begins with the placement of the subject graph, which assigns a location to every node in the graph. The subject graph typically contains many more nodes than the mapped netlist. However, as was discussed in Section 3.2.3, the area of each node in the subject graph can be normalized in order to allow the subject graph to be placed in the area designated for the mapped netlist.

The other major concern with subject graph placement is its runtime overhead, given the large number of nodes in this graph. However, this runtime can be reduced substantially by omitting the legalization of the placement. Indeed, since the congestion map is discretized at the granularity of the bins, it is reasonable to carry out the global placement of the subject graph only to the same granularity. Thus, for instance, in a recursive partitioning based placer, the placement iterations may be terminated once the size of the partitions becomes comparable to that of the bins. It is acceptable to have merely a coarse placement for the subject graph because the overlaps in this placement do not introduce larger errors in the computation of the congestion metrics than does the movement of cells during a fresh placement or legalization of the mapped netlist. Therefore, having an overlap-free placement of the subject graph does not guarantee results superior to those obtained with a coarse placement. On the other hand, carrying out the subject graph placement only to the granularity of the bins or omitting its legalization reduces the runtime for the subject graph placement significantly. The coarse placement heuristic may be particularly effective at reducing the overhead for the subject graph placement if the congestion map is also constructed on a coarse grid, as may be the case for memory efficiency reasons with the constructive congestion map based scheme discussed in Section 6.2.4.

If required, the subject graph placement may also be derived from that of the mapped netlist using some heuristics during the later iterations of the mapping, since each mapped cell can be decomposed into several primitive gates. One such heuristic assumes that all the decomposed primitives of a mapped cell are placed in the same location as the cell [LJC03].

After the placement of the subject graph, the matching phase in technology mapping proceeds in the traditional manner, as described in Section 6.1. During the matching process, each choice is assumed to be placed at the center of gravity of its fanins and fanouts as in [PB91a], in order to estimate its netlength or congestion cost or to create its constructive congestion map. Since the matching proceeds in topological order, the fanins of a node have already been processed at the time the node is processed, and the optimum matches at these fanins (and therefore, their locations) are known. The fanouts of the node, however, have still not been processed and therefore there are no matches at the fanouts yet. In order to compute the center of gravity location for the new match at the node, the locations of the fanin matches are used for the fanins, while the fanout locations are derived from the subject graph placement.

Consider the example depicted in Fig. 6.5, where the match M_p at node p in Fig. 6.5(a) is shown in Fig. 6.5(b). The location (x_{M_p}, y_{M_p}) of the match

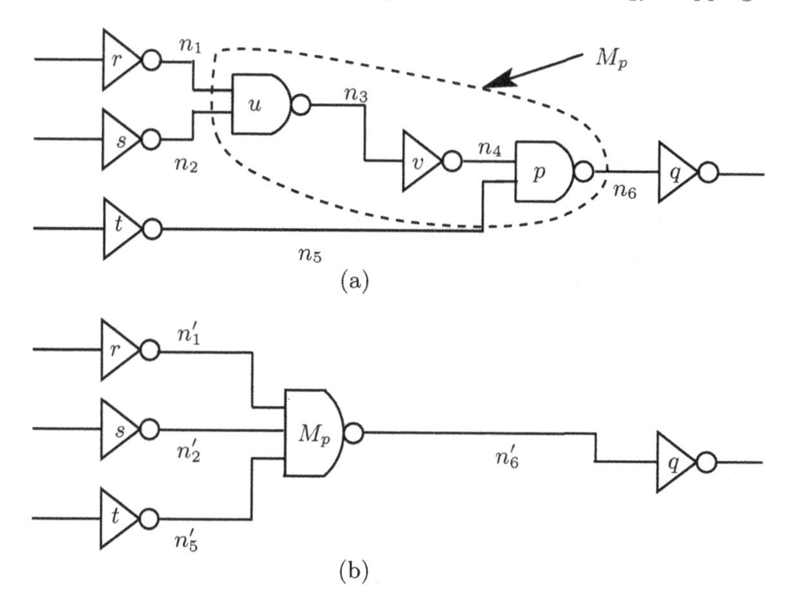

Fig. 6.5. (a) Subject graph with match M_p that subsumes the wires n_3 and n_4. (b) Placement of the match at the center of gravity of its fanins and fanouts.

is given by the following:

$$x_{M_p} = \frac{x_{M_r} + x_{M_s} + x_{M_t} + x_q}{4}, \text{ and,}$$
$$y_{M_p} = \frac{y_{M_r} + y_{M_s} + y_{M_t} + y_q}{4},$$

where M_n represents the best match at node n. Note that the location of the fanout node q is taken directly from the subject graph placement, whereas the locations of the fanin nodes are taken from the placements of the matches at the respective nodes.

The covering phases of these mapping algorithms are either conventional or they exploit the slack information to minimize the congestion. Details of specific congestion-aware technology mapping algorithms are discussed in the following sections. Sections 6.2.1 and 6.2.2 describe mapping using netlength and mutual contraction, respectively, whereas Sections 6.2.3 and 6.2.4 explain the mapping algorithms based on predictive and constructive congestion maps, respectively.

6.2.1 Technology Mapping using Netlength

The netlength metric discussed in Section 3.2.1 has been used as a proxy for routing congestion while performing technology mapping in several early approaches to congestion-aware technology mapping [PPS03, KS01]. The typical cost functions used by these algorithms are of the form:

$$K_1 \times \text{Area} + K_2 \times \text{Delay} + K_3 \times \text{Netlength},$$

where K_1, K_2, and K_3 are user-specified constants that allow different relative weights for the area, delay, and netlength of the design. At each node, the matching phase of these algorithms stores the match that minimizes the cost function being used. The covering phase, which is the same as that in traditional mapping, creates the mapping solution by selecting among these matches. The quality of the solution may be improved further by running the entire mapping procedure multiple times with different values of K_1, K_2, and K_3 to generate multiple mapping solutions, and then selecting the best of these solutions for subsequent layout.

The netlength metric has been employed as a proxy for congestion with both tree mapping and DAG mapping algorithms. The tree mapping algorithm in [PPS03] computes the total netlength of a mapping solution due to a match, whereas the DAG mapping in [KS01] employs the local wiring cost of a match instead. These computations of the total netlength and the wiring cost in these approaches are explained in the following sections.

Computation of Total Netlength During Tree Mapping

The computation of the total netlength $Nl(M_N)$ of the mapping solution due to a match M_N at node N is similar to that of area and is given by the following equation:

$$Nl(M_N) = \sum_{f \in \text{fanin}(M_N)} (Nl(M_f, M_N) + Nl(M_f)), \tag{6.5}$$

where $Nl(M_f, M_N)$ denotes the Manhattan length of the net connecting the output of the stored optimum match at fanin f and the corresponding input pin of the match M_N. In other words, $Nl(M_f, M_N)$ is given by:

$$Nl(M_f, M_N) = |x_{M_f} - x_{M_N}| + |y_{M_f} - y_{M_N}|.$$

The $Nl(M_f)$ component in Equation (6.5) represents the Manhattan lengths of all the nets in the transitive fanin cone of the mapping solution due to the match M_f at f, that have been computed previously using Equation (6.5).

Thus, using the above equations, the netlength of the mapping solution due to match M_p in the example depicted in Fig. 6.5 can be written as:

$$\begin{aligned} Nl(M_p) &= (|x_{M_p} - x_{M_r}| + |y_{M_p} - y_{M_r}|) + (|x_{M_p} - x_{M_s}| + |y_{M_p} - y_{M_s}|) \\ &+ (|x_{M_p} - x_{M_t}| + |y_{M_p} - y_{M_t}|) + Nl(M_r) + Nl(M_s) + Nl(M_t). \end{aligned}$$

The netlength cost of the match at a node is then combined with the costs for the area and the load-dependent delay (computed as described in

Section 6.1) in the cost function; the match that results in the smallest cost is stored at the node. At any given multifanout point, the overall cost (along with its area and netlength components) is divided by the number of fanouts at that multifanout point and then propagated forward; this heuristic, which is similar to the one in [CP95], allows a meaningful computation of the overall cost of the mapping solution at the primary outputs.

Computation of Wiring Cost During DAG Mapping

In the case of DAG mapping, a cost function like the total netlength of a match cannot be propagated accurately in a single topological traversal, because some of the single fanout nets from the subject graph may have to drive multiple copies of their receivers due to subsequent logic replication. For example, in Fig. 6.3(a), if both the matches M_1 and M_2 are selected during the covering, then the fanin nodes which initially have only a single fanout to u and v, respectively, now drive two receivers each. Since the actual number of fanouts is known only after the covering, the netlength propagation during the matching phase can be highly erroneous. To overcome this problem, the DAG mapping scheme proposed in [KS01] relies on the notion of the local *wiring cost* which considers the netlength of only the nets affected by the match while ignoring the effect of any subsequent logic replication during the covering phase.

For instance, the match M_p in Fig. 6.5(a) can be thought of as one that eliminates the wires n_1, n_2, n_3, n_4, n_5, and n_6, and creates new wires n'_1, n'_2, n'_5 and n'_6. Therefore, the wiring cost $WC(M_p)$ of the match can be computed as:

$$\begin{aligned} WC(M_p) = &[Nl(n'_1) + Nl(n'_2) + Nl(n'_5) + Nl(n'_6)] \\ &- [Nl(n_1) + Nl(n_2) + Nl(n_3) + Nl(n_4) + Nl(n_5) + Nl(n_6)], \end{aligned}$$

where $Nl(n_i)$ is the Manhattan netlength of the net n_i. Thus, for instance, while computing the cost of match M_p, the length of the net n_1 is given by:

$$Nl(n_1) = |x_{M_r} - x_u| + |y_{M_r} - y_u|,$$

whereas the netlength of the n'_1 is given by:

$$Nl(n'_1) = |x_{M_r} - x_{M_p}| + |y_{M_r} - y_{M_p}|,$$

with M_p assumed as being placed at the center of gravity of the location of q and those of the matches at nodes r, s, and t.

This wiring cost represents the local improvement in the wiring requirement due to the match. It is then combined with the gain-based delay of the mapping solution due to the match to determine the overall cost of the match. At any given node, the match with the least cost is stored as the

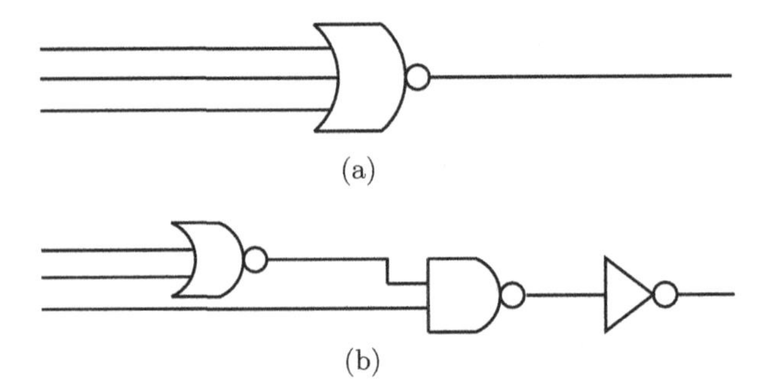

Fig. 6.6. Two different mapping solutions for the network in Fig. 6.5: solution (a) has a larger netlength and wiring cost than the one in (b).

optimal match at that node. The covering phase proceeds in the traditional manner to select among the choices stored during the matching.

Two possible solutions for the network in Fig. 6.5 are shown in Fig. 6.6(a) and (b); each cell is assumed to be placed at the center of gravity of its fanins and fanouts. One can observe that the three-input NOR implementation in Fig. 6.6(a) may have smaller cell area or cell delay than the solution in Fig. 6.6(b), but may have larger netlength. Thus, unlike classical technology mapping algorithms, those that use netlength or wiring cost during the matching phase may prefer the solution in Fig. 6.6(b).

Limitations, Time Complexity, and Extensions

As was discussed in Section 3.2.1, the netlength metric suffers from an inability to capture the spatial and locality aspects as well as the thresholded nature of routing congestion. Moreover, even though minimizing the netlength may not always lead to a reduction in congestion, it may yet result in significant area or delay penalties. These penalties can be reduced if the congested regions can be predicted accurately and the congestion-aware mapping mode applied only in these regions, with area or delay optimal choices being selected in the sparsely congested areas.

The tree mapping approach can be extended to minimize the netlength under delay or area constraints (in contrast to minimizing a linear combination of these metrics as discussed earlier in this section). This involves storing multiple non-inferior choices along with their netlengths during the matching phase. For example, consider the problem of minimizing the netlength under an area constraint using tree mapping. In this case, the matching phase can be allowed to store non-inferior choices on an area *vs.* netlength curve. Similarly, the problem of netlength minimization under delay constraints during tree mapping that uses the load-dependent delay model can be solved by keeping the track of the netlength for all the non-inferior matches stored on the load-delay curve and choosing those matches at the primary outputs that have the least netlength among all that satisfy the constraints.

The extension of DAG mapping to minimize the netlength under area or delay constraints is not obvious, as the fanouts for a net (and therefore its predicted netlength) may not be consistent across the matching and the covering phases because of logic replication.

The time complexity of congestion-aware technology mapping algorithms based on the netlength or the wiring cost is almost the same as that of conventional mapping algorithms, since the additional netlength or wiring cost computation for a match requires time that is linear in the degree of the match, which is the same as the complexity of the computation time for the area or the delay of the mapping solution due to the match. However, the runtime for the overall flow may be larger than the traditional mapping flow due to the extra overhead for the placement of the subject graph, as discussed earlier in this section.

If the mapped netlist is placed again starting from scratch, it may lose some of the netlength gains made during the technology mapping phase, since those gains are based on the companion placement of the netlist. In contrast, if the placement of the subject graph is allowed to evolve into the companion placement of the mapped netlist, which is then legalized to remove overlaps, it will not only lead to the preservation of more of the gains made during mapping, but also reduce the runtime required for the final placement.

6.2.2 Technology Mapping using Mutual Contraction

As discussed in Section 3.2.2, mutual contraction is a structural metric that does not rely on any placement information. Consequently, congestion-aware technology mapping based on this metric does not require the placement of the subject graph.

We have seen that mutual contraction is defined for two-pin nets as the product of their relative weights. For instance, using the notation introduced in Section 3.2.2, the mutual contraction $mc(n'_1)$ for net n'_1 in Fig. 6.5(b) is computed as follows (assuming unit weights for all the edges):

$$mc(n'_1) = w_r(r, M_p) \times w_r(M_p, r) = \frac{w(r, M_p)}{\sum_i w(r, i)} \times \frac{w(M_p, r)}{\sum_i w(M_p, i)} = \frac{1}{2} \times \frac{1}{4} = \frac{1}{8}.$$

The definition of mutual contraction can be extended to a match by adding the mutual contractions of all its fanins. Thus, for our example,

$$mc(M_p) = mc(n'_1) + mc(n'_2) + mc(n'_5) = \frac{1}{8} + \frac{1}{8} + \frac{1}{8} = \frac{3}{8}.$$

It may be recalled that the mutual contraction of a net correlates inversely with the expected length of the net. It has been shown that the average mutual contraction computed over all the nets in a design is negatively correlated to its Rent's exponent [LM05]. In other words, the higher the average mutual contraction for a circuit, the smaller is its Rent's exponent, resulting in reduced interconnect complexity and shorter total netlength.

Therefore, the technology mapping algorithm presented in [LM05] uses the average mutual contraction of a circuit as a proxy for its congestion. This algorithm is based on tree mapping (because the subsequent logic replication possible with DAG mapping can complicate the accurate computation of the mutual contraction of individual nets during the matching phase), and has been formulated for the optimization of the area along with the congestion. The matching phase of this algorithm assigns to each match a cost which is a linear combination of the area and the average mutual contraction of the mapping solution due to that match. The computation of the area is performed as per Equation (6.1), whereas the estimation of the average mutual contraction is carried out as explained next.

The number of nets $N_{nets}(M_N)$ in the mapping solution due to some match M_N at node N is given by:

$$N_{nets}(M_N) = |\{f : f \in \text{fanin}(M_N)\}| \quad + \sum_{f \in \text{fanin}(M_N)} N_{nets}(f),$$

where $N_{nets}(M)$ is the number of nets in the mapping solution due to a match M. Then, the average mutual contraction, $amc(M_N)$, of the mapping solution due to the match M_N can be computed as:

$$amc(M_N) = \frac{1}{N_{nets}(M_N)} \{mc(M_N) \quad + \sum_{f \in \text{fanin}(M_N)} (N_{nets}(f) \times amc(f))\}.$$

Since the average mutual contraction is negatively correlated with the total netlength, the overall cost of the mapping solution due to the match is given by:

$$\text{Cost}(M_N) = (1 - K) \times A(M_N) - K \times amc(M_N),$$

where K is a user-specified constant such that $0 \leq K \leq 1$, and $A(M_N)$ is the area of the mapping solution due to the match M_N. The match with the least cost is stored as the optimum one during the matching phase, whereas the covering phase of the algorithm is similar to that of traditional technology mapping algorithms.

Using the above equations, the number of nets in the mapping solution due to the match M_p in our example, depicted in Fig. 6.5(b), is given by:

$$N_{nets}(M_p) = N_{nets}(M_r) + N_{nets}(M_s) + N_{nets}(M_t) + 3,$$

whereas the average mutual contraction, $amc(M_p)$, of the mapping solution due to the match M_p is given by:

$$amc(M_p) = \frac{1}{N_{nets}(M_p)} \{N_{nets}(M_r) \cdot amc(M_r) + N_{nets}(M_s) \cdot amc(M_s) + N_{nets}(M_t) \cdot amc(M_t) + mc(M_p)\}.$$

The overall asymptotic time complexity of this technology mapping algorithm is slightly worse than that of traditional mapping, since the computation of the mutual contraction of a match M requires $O(\sum_{f \in \mathrm{fanin}(M)} deg(f))$ time, where $deg(f)$ is the degree of the node f. Note that this computation time cannot be subsumed by the complexity of area or delay computation for the match, which is linear in the number of fanins of the match. However, this is not a significant drawback, since in practice such a runtime penalty may be easily tolerated. More significantly, the time-consuming placement of the subject graph is not required for this approach. Therefore, the overall runtime for a flow based on this technology mapping algorithm may actually be much smaller than one that relies on placement information for nodes in the subject graph. On the other hand, the primary limitation of this approach lies in the indirect relation between mutual contraction and the routing congestion, because of which an improvement in the mutual contraction metrics for a design may not translate into commensurate gains in its congestion.

6.2.3 Technology Mapping using Predictive Congestion Maps

Figure 6.7 shows a technology mapping flow that relies on a predictive congestion map generated from the subject graph and its placement. As discussed in Section 3.2.3, the subject graph shares structural similarities with the mapped netlist. Therefore, when the same constraints are applied to the subject graph and the corresponding mapped netlist during the placement, the resulting congestion maps are likely to resemble each other [SSS+05].

A net passing through a highly congested region is more likely to be detoured than one traversing a sparsely congested area. For instance, of the two nets net_1 and net_2 depicted in Fig. 6.8, the latter net is more expensive from a routing perspective, because it passes through a more congested region. Unlike, say, the netlength based mapping algorithm discussed in Section 6.2.1, the use of a congestion map allows the mapping algorithm to treat nets in a context-sensitive manner depending on the congestion levels of the regions that they will be routed through.

Let us define the congestion cost $cc(n)$ for a net n as:

$$cc(n) = \sum_{B\,:\,c^B > c_{th} \,\wedge\, B \in \mathrm{bbox}(n)} U^B(n),$$

where c^B is the expected congestion in bin B, $U^B(n)$ is the probabilistic routing demand (or utilization) of the net n in bin B (computed as discussed in Section 2.2), and the summation is taken over all bins within the bounding box of n whose expected congestion is greater than some threshold c_{th}. This notion of the congestion cost of a net captures its context sensitivity well, since it penalizes the routing demand only in densely congested bins, and ignores it in bins where a sufficient number of tracks is available.

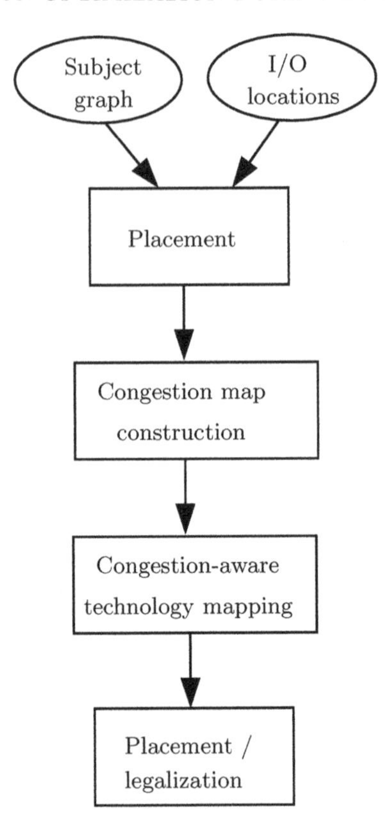

Fig. 6.7. A typical mapping flow that uses a predictive congestion map.

Figure 6.9 shows the computation of the congestion cost $cc(n)$ for a net n connecting pins p_1 and p_2, whose bounding box has sixteen bins, and each bin has a predicted congestion value associated with it. As can be seen from the figure, there are six single and double bend routes $r_1, \ldots, r_6$ possible for this net, each of which is assumed to have the same probability (namely, 1/6) of being chosen by the router. If we further assume that c_{th} is 1.0 for all the bins, then the congestion cost of the net is the sum of its routing demands in the two shaded bins (whose congestion is 1.1 and 1.2, respectively). Assume for the sake of simplicity that the number of (horizontal or vertical) tracks available in each bin is the same, and is given by N_{tr}. Since three routes (namely, r_1, r_4, and r_5) pass through the bin at location $(3, 4)$ that has a predicted congestion of 1.1, the routing demand of n in this bin is given by:

$$U^{(3,4)}(n) = \frac{1}{N_{tr}} \times \frac{3}{6}.$$

Similarly, the routing demand of this net in the bin $(3, 3)$ that has a predicted congestion of 1.2 is:

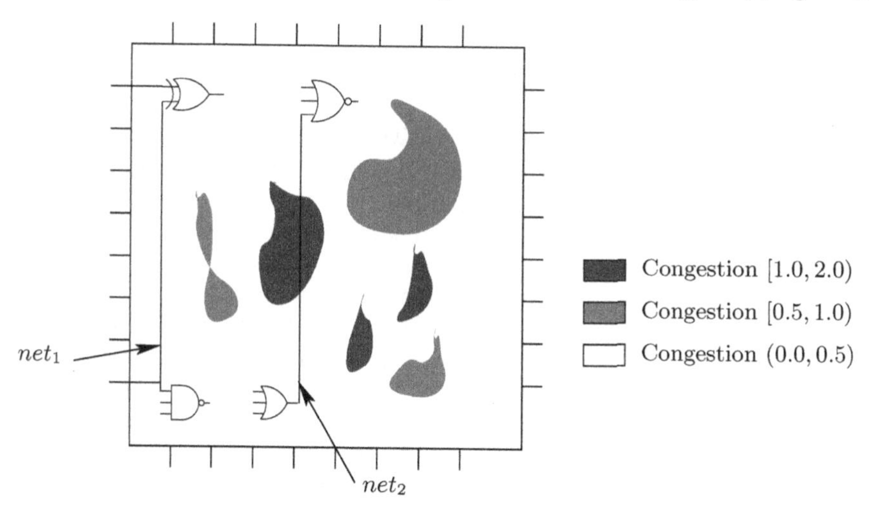

Fig. 6.8. The cost of routing a wire depends on the congestion along its route. (Reprinted from [SSS+05], ©2005 IEEE).

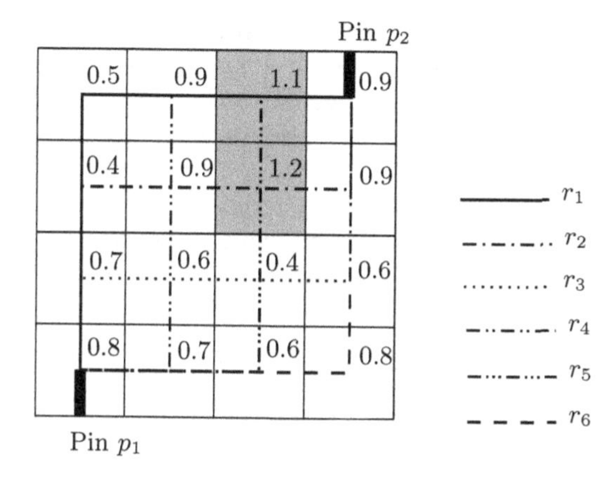

Fig. 6.9. Congestion cost of a net. (Reprinted from [SSS+05], ©2005 IEEE).

$$U^{(3,3)}(n) = \frac{1}{N_{tr}} \times \frac{2}{6}.$$

Therefore, the overall congestion cost $cc(n)$ for the net is given by:

$$cc(n) = \frac{1}{N_{tr}}(\frac{1}{2} + \frac{1}{3}).$$

The concept of the congestion cost of a net can be extended to matches also, being defined as the difference between the congestion costs of the nets created and eliminated by the match. Using this definition, the congestion cost $cc(M_p)$ for the match M_p in Fig. 6.5 can be written as:

$$cc(M_p) = (cc(n'_1) + cc(n'_2) + cc(n'_5) + cc(n'_6)) \\ -(cc(n_1) + cc(n_2) + cc(n_3) + cc(n_4) + cc(n_5) + cc(n_6)).$$

If the congestion cost of a match is positive, it implies that the nets created due to the match require tracks in densely congested regions. Conversely, a negative congestion cost of a match means that the match subsumes nets passing through densely congested regions and therefore, lowers the overall congestion. The congestion cost can be linearly combined with the area or delay cost of the match, so that the mapper will be biased towards choices that lower the congestion and improve the routability [SSS+05]. In particular, for delay optimization, the congestion cost of the match can be weighed by the slack available at that node, in order to favor the selection of matches that reduce congestion only when sufficient slack is available.

In areas that are predicted to be sparsely congested, the congestion cost of a match is zero, so that the overall cost is not unduly biased (unlike congestion-aware mapping that uses the netlength or mutual contraction metrics). In such regions, the area or delay-optimal choices are still selected, just as with traditional mapping. Thus, the mapper effectively has two modes, namely, (i) area- or delay-optimal mode in regions with sparse congestion, and (ii) congestion-aware mode in densely congested areas. These two modes enable the mapping algorithm to alleviate congestion with a relatively smaller area or delay penalty as compared to the mapping approaches discussed in Sections 6.2.1 and 6.2.2.

This mapping algorithm must pay the overhead of the placement of the subject graph, although this overhead can be reduced significantly by performing only a coarse placement and omitting legalization, as discussed earlier in this section. Furthermore, for delay optimization, there is the added runtime cost of an extra iteration of delay-optimal matching, required for the computation of the slacks. The estimation of the congestion cost for a net requires $O(b)$ time, where b is the number of bins in a given layout area. Therefore, the computation of the congestion cost of a match requires $O(\nu_m b)$ time, where ν_m is the maximum number of nets affected by a match. Since the congestion cost computation is performed for all the matches, the time complexity for the matching process is $O(mb)$, where m is the total number of matches over the entire network. In other words, the asymptotic time complexity of this mapping algorithm has an extra factor of b as compared to classical mapping.

6.2.4 Technology Mapping using Constructive Congestion Maps

As was discussed in Section 3.2.3, one of the major limitations of using a predictive congestion map is that it considers the routing demand due to all the nets in the subject graph while identifying congestion hot spots, rather than that due to only the relevant set of nets that actually exist in the mapped netlist. The constructive congestion maps described in Section 3.2.4 overcome this limitation by being created individually in a bottom-up fashion for each delay-optimal mapping solution. Moreover, the technology mapping approach that uses these maps [SSS06] also extends the covering phase to exploit the flexibility available in the form of delay slacks to reduce the congestion even further. This is in contrast to all the other congestion-aware technology mapping schemes discussed earlier in this chapter; they limit themselves to biasing the matching process only, leaving the covering phase untouched.

The following sections describe the matching and covering phases of this approach in more detail.

Matching with Constructive Congestion Maps

As is the case with other mapping approaches that use placement-dependent metrics, the matching phase in this approach assumes a prior placement of the subject graph, as well as the center of gravity placement scheme for the matches that has been described earlier. At any given node, a constructive congestion map is created for every non-inferior delay match at that node; this congestion map for the mapping solution due to that match accounts for the routing demand because of all the nets in its transitive fanin cone. As described in Section 3.2.4, this congestion map can be created incrementally for a match by adding the congestion maps for the mapping solutions due to the matches at the fanin nodes of the match to that for the fanin nets of the match.

A multipin net at a multifanout point can be modeled using star or clique[5] topologies; the star is usually preferred if the subsequent routing is going to be timing-driven.

A constructive congestion map is propagated forward across a multifanout point by dividing each congestion value in the map by the number of fanouts. In effect, the congestion caused by the nets in the transitive fanin cone of the mapping solution due to a match is distributed equally among its fanouts. This heuristic, similar to one proposed in [CP95] for area minimization under delay constraints, allows the construction of a complete congestion map for the entire mapping solution through bin-wise addition of the congestion maps corresponding to the matches selected at the primary outputs. For example, Fig. 6.10 shows the congestion map for a match that has a fanout of two. The congestion values in this map are halved during the forward propagation of the map across the multifanout point during the matching phase. Thus, the choices at nodes N_2 and N_3 (whose fanins include the net associated with node N_1) use the congestion map shown on the right.

Constructive congestion maps are stored for all the non-inferior delay matches at a node. Note that the congestion map stored at an interior node

[5] With a clique model, the congestion contributions of the resulting two-pin nets are scaled down by a factor of $2/n$, since the clique contains $n(n-1)/2$ edges, whereas only $n-1$ edges are required to connect n pins.

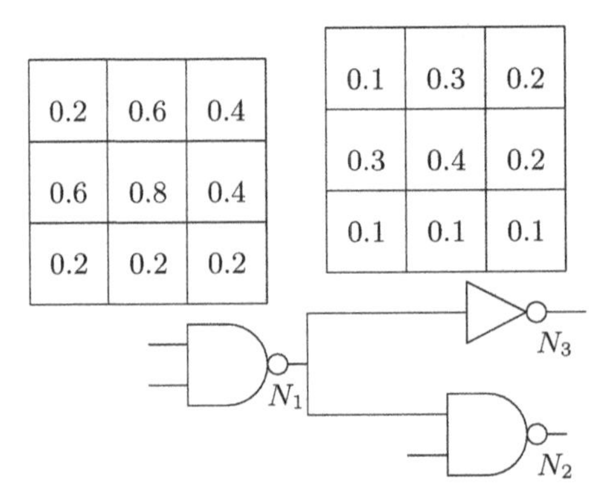

Fig. 6.10. Propagation of a congestion map across a multifanout point. (Reprinted from [SSS06], ©2006 IEEE).

in the subject graph represents only a partial picture of the congestion in the design, since it ignores the congestion contributions of nets that are not in the transitive fanin of that node. A complete congestion picture can be obtained only at the primary outputs. Therefore, the technology mapping algorithm postpones evaluating the mapping solutions based on congestion to the covering phase.

Thus, with the help of simple algebraic operations such as addition and multiplication, distinct two-dimensional congestion maps can be constructed and propagated forward for different mapping solutions during the matching phase. These maps are utilized during the covering to decide a mapping solution which reduces congestion.

Exploiting Slacks during the Covering Phase

During the traditional delay-oriented covering process, the match chosen at a given node is the one that minimizes the delay for the (known) load at that node. The load-delay curves constructed during the matching phase also assume the same goal of delay minimization. However, observe that it is possible to choose a suboptimal match at a node with positive slack and yet not violate any delay constraints. The covering algorithm proposed in [SSS06] employs this idea to minimize the congestion, while still guaranteeing delay optimality.

Consider the load-delay curve shown in Fig. 6.11, which has been constructed at some node during the matching phase. When the node is processed during the covering phase, let us assume that it has a slack of 10 units and has to drive a load of 15 units. The delays due to matches M_1, M_2, and M_3 for this load are 95, 90, and 95 units, respectively. In this case, regular covering selects match M_2, since it minimizes the delay. However, in order to reduce the congestion, the covering should choose the match that minimizes the congestion, as long as it does not violate the delay constraints. The choice

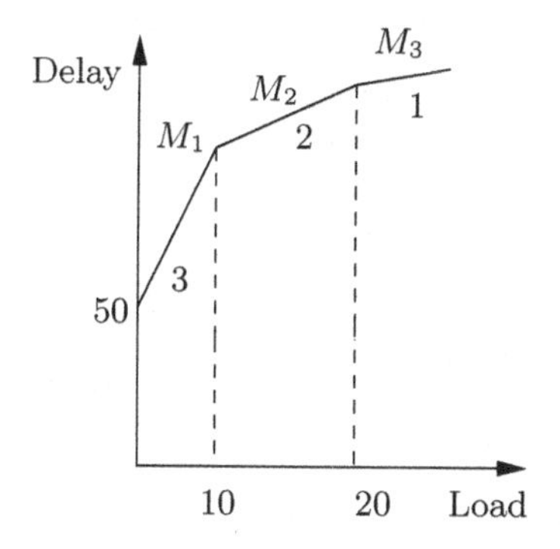

Fig. 6.11. A piecewise linear load-delay curve with three matches. (Reprinted from [SSS06], ©2006 IEEE).

of either M_1 or M_3 does not affect the delay-optimality of the overall solution, because there is a slack of 10 units available at the node and the arrival times due to M_1 and M_3 both satisfy this slack constraint.

Algorithm 14 Perform congestion-aware covering under delay constraints (Reprinted from [SSS06], ©2006 IEEE)

Input: A Boolean network $G(V, E)$, a set of primary outputs $O \subseteq V$, sets of non-inferior matches M_v and their congestion maps CM_v ($\forall v \in V$)

Output: Assignment of congestion-optimal matches $m_v^{opt} \in M_v$ which satisfy the delay constraints ($\forall v \in V$)

1: **for all** $o_i \in O$ **do**
2: $\quad m_{o_i}^{Dopt} \leftarrow \text{DelayOptimalMatch}(M_{o_i}, load_{o_i})$
3: $\quad s_{o_i} \leftarrow D_{o_i}^{req} - D_{m_{o_i}^{Dopt}}$
4: **end for**
5: $CM \leftarrow \sum_{i=1}^{|O|} CM_{m_{o_i}^{Dopt}}$
6: $OF \leftarrow \text{ComputeOverflow}(CM)$
7: **for all** $v \in V$, in reverse topological order, **do**
8: $\quad m_v^{Dopt} \leftarrow \text{DelayOptimalMatch}(M_v, load_v)$
9: $\quad m_v^{Copt} \leftarrow \text{CongestionOptimalMatch}(M_v, s_v)$
10: $\quad$ **if** $OF_v^{Copt} < OF$ **then**
11: $\quad\quad m_v^{opt} \leftarrow m_v^{Copt}$
12: $\quad\quad CM \leftarrow CM - CM_{m_v^{Dopt}} + CM_{m_v^{Copt}}$
13: $\quad\quad OF \leftarrow OF_v^{Copt}$
14: $\quad\quad \text{UpdateSlacks}(m_v^{Copt}, s_v^{Copt})$
15: $\quad$ **else**
16: $\quad\quad m_v^{opt} \leftarrow m_v^{Dopt}$
17: $\quad\quad \text{UpdateSlacks}(m_v^{Dopt}, s_v)$
18: $\quad$ **end if**
19: $\quad \text{UpdateLoads}(m_v^{opt})$
20: **end for**

The pseudocode for the covering that targets routing congestion under delay constraints is shown in Algorithm 14. It begins with the computation of the delay-optimal matches at the primary outputs, followed by the estimation of slacks for all the outputs. The congestion map CM for this solution is built by the bin-wise addition of the congestion maps due to delay-optimal matches for all the primary outputs. The total track overflow OF corresponding to this solution is estimated from the congestion map by adding the track overflow values in all congested bins.

After this initialization, all the nodes ($v \in V$) are processed in the reverse topological order. First, for any given node v, the delay-optimal match m_v^{Dopt} is determined for that node, followed by the computation of match m_v^{Copt} that satisfies the slack constraint and has the least congestion cost among all the matches stored on the load-delay curve at that node. This match is selected and the resulting slack and congestion map updates are propagated to the nodes at the fanins of the match.

Optimality, Time Complexity, and Extensions

With the use of constructive congestion maps, the mapping algorithm discussed in this section can ensure the delay optimality of a mapping solution (in contrast to the other congestion-aware technology mapping algorithms proposed to date, that do not make any claims of delay optimality). However, as with the other algorithms, it does not ensure the optimality of the track overflow or the peak congestion.

This mapping algorithm requires $O(mb)$ additional time over conventional mappers, where b is the number of bins and m is the total number of matches over the entire network. As with the mapping approach that uses predictive congestion maps, the actual runtimes for this approach are usually a few times that of the corresponding runtimes for classical delay-oriented technology mapping in practice.

The memory requirement of technology mapping based on constructive congestion maps can be significantly larger than that for conventional mappers, due to the storage of the congestion maps for all the non-inferior choices preserved during the matching phase. This requirement may be reduced substantially, for instance, by storing only those bins that lie within the region affected by a mapping solution, or by using a coarse grid. The use of a coarse grid for the congestion map can also help reduce the runtime overhead for the subject graph placement substantially, if that placement is carried out only to the granularity of this grid (as described at the beginning of Section 6.2). Another option for improved memory efficiency is to store congestion maps only along the "wavefront" of the nodes being processed at any time, as in [SIS99].

The extension of this mapping algorithm to DAG mapping is not obvious because of the duplication of nets corresponding to replicated logic.

6.2.5 Comparison Of Congestion-aware Technology Mapping Techniques

In this section, we discussed four different congestion-aware technology mapping techniques, each of which uses a different metric. The comparison of these techniques with respect to various criteria is reviewed next.

Time Complexity and Runtime : The asymptotic complexity of the technology mapping approaches that use the netlength as a metric for congestion is almost the same as that of the conventional mapping algorithm. However, the overall runtime for the design flow may suffer due to the overhead for the placement of the subject graph (although this overhead may be reduced significantly using heuristics discussed below). Technology mapping based on the mutual contraction does not require such a placement and therefore its overall runtime is usually comparable to conventional mapping, even though the asymptotic time complexity of the matching phase is slightly worse. The complexity of the mappers that use congestion maps is affected by $O(mb)$ additional computation time as compared to conventional mappers. This overhead affects only the matching phase in the mapper based on a predictive map, but affects both matching and covering phases in the mapper that uses constructive congestion maps. Moreover, these congestion map based approaches also require the overhead of subject graph placement. Their runtime is typically a few times that of conventional mappers.

Memory Complexity : The memory requirement of the mapping approaches based on netlength, mutual contraction, and predictive congestion maps is almost the same as that of conventional mapping. In contrast, technology mapping based on constructive congestion maps requires $O(mb)$ memory with a naïve implementation, although this memory overhead can be substantially reduced through various memory-efficient implementation techniques and heuristics.

Placement of the Subject Graph : The subject graph must be placed for all the mapping approaches discussed here except for the one based on mutual contraction. This placement can claim a significant share of the runtime, since the size of the subject graph is usually much larger than that of the corresponding mapped netlist. However, this overhead can be reduced significantly by performing a coarse placement merely to the granularity of the bins, or even by omitting the legalization of the placement of the subject graph (as discussed in the preamble to Section 6.2). Moreover, this placement also helps improve the modeling of the parasitics and delays of the nets in the design. Indeed, most of the modern physical synthesis tools rely on some placement information to model the nets during delay-oriented technology mapping.

Effectiveness and Sources of Errors : Congestion-aware mapping algorithms are affected by several sources of errors. For example, the ones based on the netlength suffer from the limitation that the netlength is an indirect global metric and may not capture the impact of congestion on the routability. Moreover, under this approach, mapping choices are penalized equally in both densely and sparsely congested regions. Mismatches between the placements of the subject graph and the mapped netlist also impact the accuracy of the netlength computation. The mapping based on mutual contraction is affected by the imperfect correlation between the mutual contraction and the routing congestion, since this metric is a proxy for the netlength, which is itself only an indirect measure of the congestion. The use of congestion maps allows the mapper to operate in congestion-aware mode only in regions that are expected to be congested, thus significantly reducing the area and delay penalties for the congestion optimization. Constructive congestion maps are usually more accurate than predictive ones, although the discrepancies between the placements of the subject graph and the mapped netlist can affect the accuracy of all congestion maps. However, if the placement assumed during the mapping is preserved by using legalization to obtain the placement of the mapped netlist, the error due to the placement mismatches can be reduced substantially. Additionally, the constructive map based technique maintains the delay optimality of the mapping solution (unlike the other techniques discussed in this section).

6.3 Overview of Classical Logic Synthesis

Logic synthesis optimizations, which have traditionally aimed at minimizing the number of literals or the number of levels in a multilevel Boolean network, employ several transformations that are based either on algebraic or on Boolean methods. Some of these include decomposition, extraction, substitution, and elimination; these operations are described in detail in [BHS90, DeM94]. There have been several attempts to extend these operations to consider congestion metrics such as netlength, adhesion, or the average neighborhood population, which were discussed in Section 3.3 of Chapter 3. From the perspective of congestion alleviation, the key to the efficacy of these operations lies (i) in their ability to identify and subsume "bad" nets and possibly replace them with "good" ones, and, (ii) in preserving the resulting gains through technology mapping and placement. The effectiveness of the former depends on the employed metric and the underlying algorithm, whereas the preservation of the gains is largely dependent on the remainder of the design flow.

In this section, we review the conventional algorithms for technology decomposition as well as those for multilevel logic synthesis, since these algorithms have been extended to consider routing congestion; Section 6.4 discusses these extensions. In this chapter, we will not describe sequential optimizations, two-level sum-of-products (SOP) expression minimization methods, or Boolean operations aimed at optimizing combinational logic, since these algorithms have not yet been explored from a congestion point of view; any standard textbook on logic synthesis (such as [DeM94]) covers them well.

6.3.1 Technology Decomposition

The technology decomposition step, which precedes the technology mapping, converts a Boolean network into a subject graph containing only primitive gates. The choice of the primitive gates, which are usually one or more of EXORs, NANDs, NORs, and inverters, has a large impact on the quality of results obtained after the technology mapping. Finding a subject graph that yields optimum technology mapping results for a given network and library has long been an open problem [BHS90]. In practice, Boolean networks are often decomposed into subject graphs composed of two-input NANDs and inverters [BHS90, SSL+92].

The Boolean function at a node in the network is often written either in SOP form or using a factored form. The former has historically been more widespread and represents an AND-OR cover for the function, whereas the latter form can lead directly to a static CMOS realization. Given a function expressed in one of these forms, it can be easily transformed into the other form. Each product term in a SOP expression is also referred to as a *cube*. The technology decomposition stage typically assumes Boolean functions to be in SOP form and decomposes each node into a tree whose nodes represent only the two-input NAND and inverter functionalities. This is facilitated by the straightforward correspondence between an AND-OR cover and the equivalent NAND-NAND network. The resulting multiple input NAND gates can be decomposed further into two-input NANDs followed by inverters.

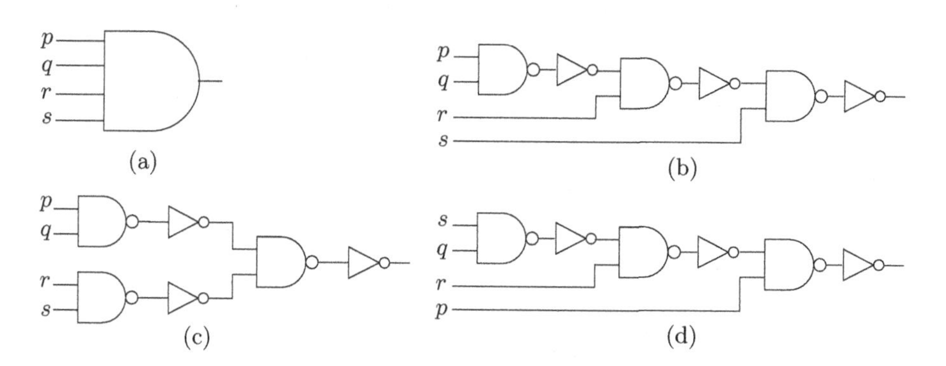

Fig. 6.12. Various technology decompositions of a four-input AND gate.

A few different technology decompositions for a four-input AND gate are shown in Fig. 6.12. These decompositions are usually created using a greedy algorithm similar to the one used for Huffman encoding. This algorithm returns optimal results for a cost function such as delay; at this technology-independent level of abstraction, the delay is often measured using the unit delay model that assumes a delay of one unit for each gate (and zero delay for the wires). Among the decompositions shown in the figure, the decomposition in Fig. 6.12(c) is delay-optimal when all the inputs arrive at the same time, whereas the ones in Fig. 6.12(b) and Fig. 6.12(d), although topologically similar, are delay-optimal when s and p, respectively, are the latest arriving inputs.

6.3.2 Multilevel Logic Synthesis Operations

The technology decomposition stage is traditionally preceded by algebraic operations that are iteratively applied on a multilevel Boolean network to optimize objectives such as the literal count or the number of logic levels. A node in the multilevel network represents a Boolean function computed on its inputs, whereas a (directed) edge denotes the input/output relationship between two nodes. Typical algebraic operations re-express the functions by eliminating some existing nodes or edges or by adding new ones, thus affecting the structure of the network. Some of the widely used transformations such as decomposition, extraction, substitution, and elimination are briefly explained below.

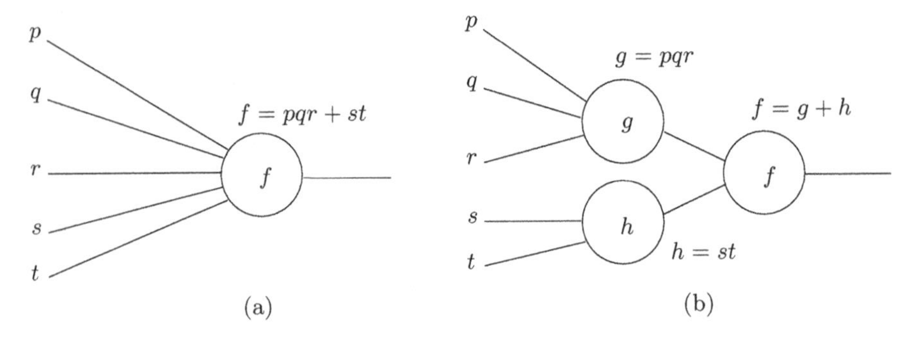

Fig. 6.13. *Decomposition* of the function f shown in (a) in terms of new variables g and h as shown in (b).

Definition 6.1. *The operation of expressing a given Boolean function in terms of new intermediate variables is known as* decomposition.

Consider a function $f = pqr + st$. In a multilevel Boolean network representation, it is represented by a node with five inputs as shown in Fig. 6.13(a). It can also be written as a disjunction of two functions $g = pqr$ and $h = st$, which introduces two new nodes, g and h, in the network, as shown in Fig. 6.13(b). The second representation is referred to as the decomposition of the original expression of f into the functions g and h. Note that this decomposition is different from the technology decomposition discussed in Section 6.3.1, since it

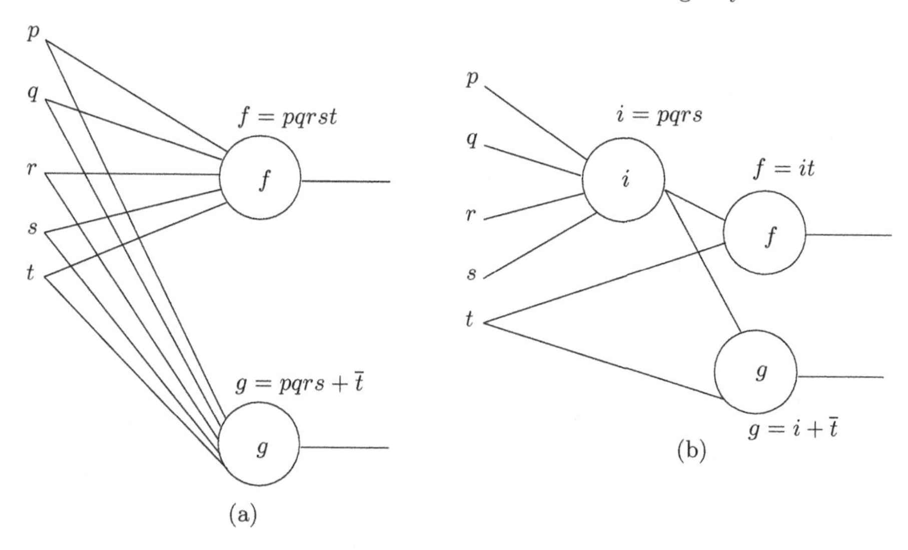

Fig. 6.14. *Extraction* of the common subexpression $pqrs$ from the functions f and g in (a), resulting in the network in (b). The network in (b) can be transformed back to the one in (a) by *eliminating* the node i.

allows arbitrary functions for the newly introduced nodes, rather than merely the primitive ones.

Definition 6.2. *The operation of expressing a set of Boolean functions in terms of old and new intermediate variables is known as* extraction.

As an example, consider the two functions $f = pqrst$ and $g = pqrs + \bar{t}$ represented by the network in Fig. 6.14(a). A common expression $pqrs$ can be extracted from both the functions and implemented separately, so that the resulting network, shown in Fig. 6.14(b), has two fewer literals than the network in Fig. 6.14(a).

Definition 6.3. *The operation of expressing the function associated with a node in terms of other nodes in the network is known as* substitution *(or* resubstitution*).*

To illustrate this transformation, assume that two nodes existing in the network represent the functions $f = pr + ps + qr + qs + st$ and $g = p + q$, respectively. Then, the substitution of g into f leads to the re-expression of f as $g(r + s) + st$.

Definition 6.4. *The operation of eliminating a given node from the entire network is known as* elimination *(or* collapse*).*

The elimination transformation can be thought of as the inverse of the substitution operation. It re-expresses the function in terms of the fanins of the nodes to be eliminated, which are then removed from the network. Thus, for instance, the network in Fig. 6.14(a) can be obtained by eliminating the node i from the network in Fig. 6.14(b).

Underlying these algebraic operations is the concept of *division* that allows the expression of a function f as $pq+r$, where p is referred to as the *divisor*, q as the *quotient*, and r as the *remainder*. For example, in the case of extraction or decomposition, if the divisors of a function are known, then the function may be expressed to reduce the number of literals. Similarly, for the elimination and substitution operations, determining whether a given node is a divisor enables the re-expression of the nodes based on the literal savings.

Primary divisors that cannot be divided evenly (*i.e.*, without a remainder) any further are known as *kernels*; kernels can be found by solving the rectangle covering problem [BHS90]. This operation can be computationally expensive if an arbitrary number of cubes is allowed in the kernels. Instead, if the search space for the divisors is restricted to at most two cubes or single cubes with two literals each, the divisors can be found in polynomial time [RV92]. For example, in order to find the two-cube divisors of a function with n cubes, one has to consider $n(n-1)/2$ pairs of cubes and check whether the intersection of the literals in the pairs is empty. The *fast extraction* procedure, available in the widely used sequential synthesis package SIS [SSL+92], is based on these algorithms; its pseudocode is shown in Algorithm 15. It begins with the generation of two-cube and two-literal single cube divisors. Each divisor d_i has a gain associated with it that represents the improvement in the cost of the network if the divisor is substituted into the network. For example, if the cost of the network is defined as the number of literals, the gain g_i due to a given divisor d_i is given by:

$$g_i = n(l-1) - l,$$

where n is the number of nodes in which d_i appears, and l is the number of literals in d_i. In this equation, the first term represents the literal savings due to the substitution of the divisor, whereas the second one denotes the cost of implementing the divisor. The divisor with the maximum gain is selected and substituted in the network. As a result, the gains due to other divisors may change; some divisors may even cease to exist. Therefore, the gains of all the remaining divisors are updated and the procedure is repeated until no more improvement in the cost of the network is possible.

The substitution, elimination, and decomposition operation can also be applied iteratively. For example, applying substitution requires checking whether a node can be expressed in terms of any other nodes in the network and if so, whether the transformation leads to a reduction in the number of literals. The elimination operation is performed similarly by evaluating whether the collapse of a node can lead to an improvement in the cost of the network.

Algorithm 15 Fast extraction

Input: A Boolean network $N(V, E)$ with given cost C_1
Output: An equivalent Boolean network with cost $C_2 < C_1$
1: Find all two-cube and two-literal single cube divisors d_i
2: Associate gain g_i with each divisor d_i
3: **while** $\exists\ d_i$ with $g_i > 0$ **do**
4: Choose a divisor with the best gain
5: Substitute the divisor d_i into the network
6: Update the gain of remaining divisors
7: **end while**

6.4 Congestion-aware Logic Synthesis

Congestion-aware logic synthesis techniques try to steer the output of the technology decomposition and multilevel synthesis operations towards regions of the design space that are more friendly from a routing perspective, as defined by some metric correlated with the eventual congestion of the design. Specifically, technology decomposition can be guided by metrics such as netlength or mutual contraction, whereas other multilevel logic synthesis operations can rely on these and other structural metrics. As discussed in Chapter 3, attempts to use placement-dependent congestion metrics at the logic synthesis stage have not been very successful to date. Instead, simple graph theoretic metrics that attempt to capture the local or global connectivity of the network have been used during the logic synthesis stage as proxies for the post-routing congestion of the circuit.

The congestion-aware extensions of the technology decomposition and multilevel synthesis operations are described next, along with their limitations.

6.4.1 Technology Decomposition Targeting Netlength and Mutual Contraction

The algorithm for conventional technology decomposition, that was described in Section 6.3.1, has been extended in [KS02] and [LM05] to minimize the netlength and mutual contraction metrics, respectively. The pseudocode for such a congestion-aware technology decomposition is shown in Algorithm 16. In each iteration of the while loop, the algorithm reduces the pin-count of the gate G, which is being decomposed, by one. This is achieved by finding a pin pair with the least netlength or the highest mutual contraction between the two pins, and creating a new two-input NAND gate to which the selected pin pair serves as input. The output of this two-input gate is connected to an odd number of inverters in order to preserve the logic functionality; the output of the last inverter becomes an input to G. Thus, in each iteration, the pin count of G is reduced by one. Therefore, $n - 1$ iterations are required to convert an n-input gate into a tree of two-input gates and inverters.

Algorithm 16 Congestion-aware technology decomposition minimizing netlength or mutual contraction

Input: A NAND gate G with n inputs
Output: Decomposed network containing two-input NAND gates and inverters
1: **while** number of input pins to $G > 2$ **do**
2: Compute the distance or mutual contraction between all pin pairs
3: $(p1, p2) \leftarrow$ Pin pair with the least distance or the highest mutual contraction
4: Create two-input NAND with (p1, p2), add an odd number of inverters at its output, and place these new gates
5: Remove (p1, p2) from the input pins of G
6: Add the output of the last inverter from Step 4 as an input to G
7: **end while**

The algorithm can easily be made timing-aware by considering only those pin pairs whose signal arrival times differ by no more than a given bound. Although the pseudocode presents the decomposition of a multiple input NAND node into a tree of two-input NANDs and inverters, the decomposition of nodes with other functionalities and into different primitive gates such as NORs and EXORs can also be carried out in a similar fashion. The optimality of the solutions employing this greedy algorithm, however, cannot be ensured for either of the two congestion-based cost functions. In the case of decomposition aimed at optimizing the netlength metric, the two-input gates and inverters created by decomposition are usually assumed to be placed in the center of gravity of their fanins and fanouts even though many different placements (that lead to different netlengths) are possible, leading to a potentially inaccurate netlength estimation. Moreover, the decisions made in the first few iterations based on the center of gravity placements affect the choices of decompositions in later iterations. A similar argument applies to the decomposition aimed at maximizing the mutual contraction. An exact algorithm based on branch and bound technique may be used, but the runtime of such an algorithm tends to be high, since in the worst case, $n!(n-1)!/2^{(n-1)}$ decompositions may have to be considered for an n-input node.

In general, it is observed that a technology decomposition algorithm that considers the netlength or mutual contraction metrics often generates better decompositions than one that is completely oblivious to the layout. However, even with placement awareness, a technology decomposition algorithm may make suboptimal choices if it does not comprehend congestion. As an example, consider Fig. 6.15 that shows two different decompositions of a four-input AND gate. In this case, the input drivers and the output receiver are assumed to have fixed locations. Figure 6.15(a) shows a decomposition that can be obtained by a greedy algorithm that optimizes the netlength metric. This decomposition may lead to a technology mapping solution with small delays when the regions in the neighborhoods of the nets in the solution are sparsely congested. However, in the presence of congested regions (such as the shaded area in the figure), greedily choosing the decomposition that minimizes the

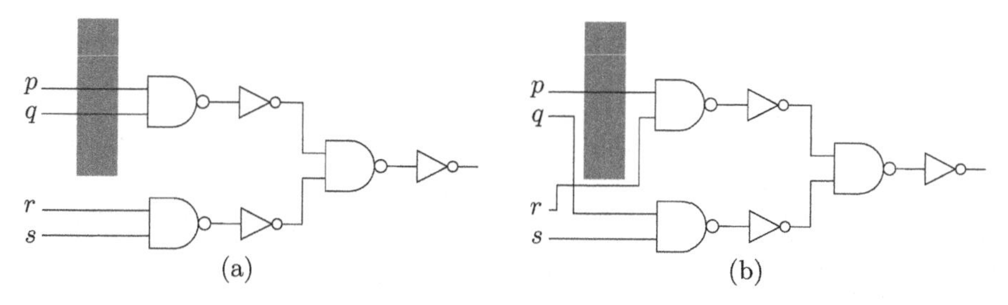

Fig. 6.15. Two technology decompositions of a four-input AND gate: (a) Decomposition with minimum netlength based on Manhattan distance. (b) Congestion-aware decomposition that can lead to a mapping solution with better delay.

netlength purely based on Manhattan distance may lead to poor mapping solutions. Let us assume that there is only one horizontal track available in the congested area, which will result in one of the two input nets p and q being detoured (resulting in a delay penalty) if the decomposition in Fig. 6.15(a) is chosen. Moreover, the greedy choice prevents the pairing of p with r and q with s, even if it can lead to a mapping solution with small delays. On the other hand, a truly congestion-aware decomposition that comprehends the congested region and its impact on the netlength and the delay can pair p with r and q with s, as shown in Fig. 6.15(b). Although this leads to a slightly larger netlength based on Manhattan distance, it is likely to result in smaller post-routing wirelengths and delays as compared to the decomposition in Fig. 6.15(a).

6.4.2 Multilevel Synthesis Operations Targeting Congestion

One can observe that multilevel synthesis operations affect the structure of the network, as new nodes and edges are introduced and old ones are removed. For instance, the decomposition operation shown in Fig. 6.13(b) adds two new nets, represented by nodes g and h. This is desirable if these nets are likely to be short and also reduce the lengths of the remaining five nets that exist in the equivalent undecomposed network in Fig. 6.13(a). On the other hand, this decomposition may be unacceptable if the newly created nets are long or have to traverse congested regions. Thus, whether a given logic synthesis transformation is good or bad partly depends on the placement context. In congestion-aware logic synthesis, this evaluation and the subsequent decision to accept or reject a transformation is guided by the synthesis-level metrics discussed in Chapter 3.

A multilevel synthesis operation can be made "congestion-aware" by augmenting its traditional cost function (often, the literal count reduction) with an additional congestion cost (as measured by one of the synthesis level congestion metrics). As a result, the congestion metric will influence which divisors are chosen or filtered out during the divisor extraction process; however, the underlying algorithms to extract divisors or substitute them into the network remain unchanged.

Most of the multilevel synthesis approaches targeting congestion fall into the category described above, except for an early work described in [SA93]. That work, on multilevel synthesis targeting routing area, does not use any metric to guide kernel extraction. Instead, it relies on the notion of a lexicographical order to filter out the divisors that are not compatible with the given order of inputs. Although this approach showed promising results in a 1.2 μ technology, it is not clear how effective such an approach will be in modern technologies, in which the typical design sizes are large, multiple metal layers are available for routing of input and intermediate signals, and the nets connected to the input pins contribute only a small fraction of the total wirelength of the network.

A design flow[6] for congestion-aware multilevel logic synthesis optimizations that uses the netlength as the congestion metric is shown in Fig. 6.16. It begins with the creation of a subject graph placement for all the nodes in the Boolean network. If the placement is created directly from the Boolean network, each node is assigned an area, which is either the same for all nodes or is proportional to the number of literals in the function associated with the node. In subsequent design iterations, the subject graph placement may also be derived from the companion placement of the mapped netlist with the help of some heuristics (such as the one in [LJC03] that places all the decomposed primitives of a mapped cell in the same location as the cell). It is not necessary to eliminate all overlaps from the subject graph placement, since they do not affect the quality of results significantly, as discussed in Section 6.2. This placement can help evaluate the netlength or the change in total netlength of the network due to the substitution of a given divisor. For example, the netlength cost $C_{Nl}(i)$ for the divisor i in Fig. 6.14(b) is merely the sum of the lengths of the edges connecting i with its fanins and fanouts, as shown in the following equation:

$$C_{Nl}(i) = Nl(p,i) + Nl(q,i) + Nl(r,i) + Nl(s,i) + Nl(f,i) + Nl(g,i), \quad (6.6)$$

where $Nl(x, y)$ is the netlength metric that measures the Manhattan distance between two nodes x and y. This netlength cost of the divisor is used in [KS02] to select the divisor with the least length from a set of divisors that result in approximately equal literal savings. On the other hand, the kernel extraction proposed in [PB91b] relies on the change in the total netlength as a proxy for

[6] The wireplanning work in [GNB+98, GKS01] uses a similar flow with interleaved synthesis and placement, but it does not specifically target the congestion problem. It applies filtering to the selection of divisors by either discarding the ones that do not have good placements that can minimize the netlengths on the paths between inputs and outputs, or by duplicating kernels and placing them such that the paths have minimum netlengths.

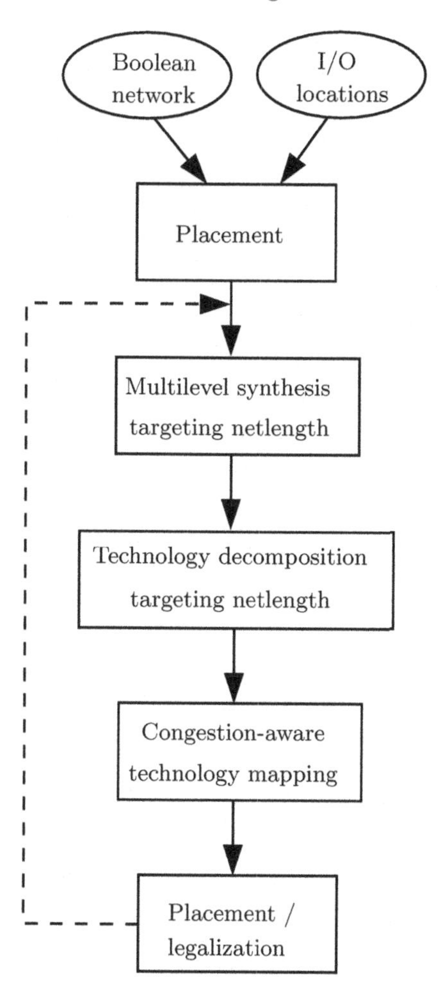

Fig. 6.16. The multilevel synthesis operations that use netlength as a metric require the placement of a Boolean network.

the routing cost and the congestion while choosing a divisor that can maximize the literal savings. The change in the total netlength of the network, Δ_{TNL}, can be computed by considering the nets that are affected by the divisor that is being substituted. For example, the substitution of the divisor i in Fig. 6.14 removes the edges between the input nodes p, q, r, and s and the internal nodes f and g, and adds new edges between the same input nodes and the new node i, and between the nodes i and f as well as i and g. Therefore, the node i results in the following change in netlength:

$$\begin{aligned}\Delta_{TNL}(i) = \{&Nl(p,f) + Nl(q,f) + Nl(r,f) + Nl(s,f) \\ &+ Nl(p,g) + Nl(q,g) + Nl(r,g) + Nl(s,g)\} \\ &- \{Nl(p,i) + Nl(q,i) + Nl(r,i) + Nl(s,i) + Nl(i,f) + Nl(i,g)\}.\end{aligned} \tag{6.7}$$

While carrying out the computation in Equations (6.6) and (6.7), the divisor is assumed to be placed at the center of gravity of the locations of the fanins and the fanouts of the new node. This is then used for congestion-aware fast extraction in [KS02] and for arbitrary divisor extraction and elimination in [PB91b].

Some recent work has shown that the optimum placement for a divisor that is likely to result in the least increase in netlength lies inside the rectangle bounded by the two horizontal (and the two vertical) medians obtained from the sets of the edges of the bounding boxes of the fanins and the fanouts, respectively, sorted by their x (y) coordinates [CB04]. Any location within this rectangle will result in the minimum change in the total netlength of the network. This placement is applied to the fast extraction of congestion-aware divisors. In this framework, the gain G_i due to a divisor d_i is given by:

$$G_i = \lambda g_i + (1 - \lambda) Nl_i,$$

where $0 \leq \lambda \leq 1$ determines the relative weights for the literal savings g_i and the netlength gain Nl_i. The netlength gain used here can be defined using either Equation (6.6) or (6.7). In each iteration of the fast extraction, a node with the best gain is selected, as shown in Algorithm 15, and the gains of the remaining divisors are updated.

Multilevel synthesis operations that use graph theoretic metrics such as the average neighborhood population and overlaps in fanout ranges to improve the routability do not require the placement of a Boolean network [KSD03, KK04, VP95]. The concepts of the fanout range and the neighborhood population have been discussed in Sections 3.3.3 and 3.3.4, respectively. A node that has a large average neighborhood population at small distances is likely to have denser connectivity or entanglement as compared to other nodes that have a low population. Regardless of the placement, such high population nodes are likely to result in many connections competing for routing resources in the same areas, thus causing congested hot spots. The goal of the congestion-aware versions of algebraic operations such as elimination, resubstitution, and speed-up[7] in [KSD03] and Boolean restructuring in [KK04]

[7] The *speed-up* operation in multilevel synthesis aims to improve the overall delay of the network [BHS90]. It involves the following steps: (i) determining the timing-critical paths in the network, (ii) forming a cut-set of nodes such that reducing the arrival times at these nodes improves the overall delay of the circuit, (iii) applying elimination operation to the nodes so that they are expressed in terms of primary inputs or other intermediate variables, and (iv) performing timing-driven technology decomposition on the nodes to improve the arrival times.

is to minimize the average neighborhood population at smaller distances. Intuitively, it means that the network with many short nets is likely to have improved routability and is, therefore, desirable. The work in [KK03] empirically demonstrates that algebraic operations such as elimination and extraction can indeed have a significant impact on the average neighborhood population. For instance, the elimination operation, that expresses a node in terms of the fanins of the eliminated nodes, usually increases the neighborhood population at small distances, but reduces the overall population. On the other hand, the extraction operation, which substitutes common divisors to reduce the literal count, often reduces the average neighborhood population at small distances but increases the number of nodes and therefore, the average population at large distances.

It has been conjectured [VP95] that reducing the overlap in the fanout intervals over all nets in the network improves the routability, where the fanout interval corresponds to the fanout range of the net. This work uses the function $\sum_{ov} n_{ov}^2 l_{ov}$ as a minimization objective during the logic extraction, where n_{ov} denotes the number of overlapping fanout intervals having overlap length l_{ov}, and the summation is taken over all the overlaps in the fanout intervals. Minimizing this cost function implies that extractions that reduce the overlaps and the fanout range are favored in the likelihood that this will create many (short) nets that are mostly restricted to local regions. One of the problems with this objective function is that fanout range is a poor metric for predicting the netlengths of two-pin nets (since the length of the overlap is always zero for such nets), and such nets may constitute a non-trivial fraction of the total wirelength in the network.

6.4.3 Comparison of Congestion-aware Logic Synthesis Techniques

Almost all multilevel logic synthesis operations that target congestion keep the underlying traditional algorithms intact and modify the objective function by adding a congestion cost, measured using some metric such as the netlength, fanout range, or average neighborhood population, to the literal saving or the number of levels. A candidate operation, such as extraction, resubstitution or elimination, is evaluated for the change in the value of the modified objective, and the change is either accepted or rejected. Additional computation is required for the estimation of the congestion metric; this cost may be subsumed by the asymptotic cost of the evaluation of the conventional cost within the operation, or it may result in an extra term proportional to the degree of the node being processed. Moreover, approaches that use the netlength as a metric require the placement of the Boolean network, which may result in additional runtime penalties. The lexicographical extraction proposed in [SA93] does not rely on any congestion metric, and merely filters out the divisors that are not compatible with the given literal order; this check requires time that is linear in the number of literals in the divisor and can be subsumed by the time complexity of the generation of the divisor itself.

The multilevel synthesis transformations discussed in this section affect the structure of the network and improve the routability as measured by the employed metric. However, whether the transformation has any effect on the final result can be known only after the technology (re-)mapping and the subsequent placement or legalization. The congestion gains obtained during logic synthesis may not be preserved because the congestion metrics used at this stage are not very accurate; finding a congestion metric with good fidelity and computability at the logic synthesis stage remains an open research problem. Furthermore, even though the logic synthesis transformations are powerful, they can turn out to be ineffective if the subsequent steps fail to preserve (or improve) on their gains.

Indeed, because of this second reason, it is necessary for technology decomposition and mapping to be layout-aware, and incorporate congestion metrics that correlate with the one that drives the preceding multilevel synthesis operations. Of course, the holy grail of congestion-aware logic synthesis algorithms is the ability to create a structurally superior network that preserves the routability gains independent of the subsequent decomposition and mapping algorithms. Much of the research in this area has attempted to demonstrate the superiority of networks obtained by modifying the cost functions in otherwise traditional logic synthesis algorithms. Unfortunately, the empirical data and conclusions in these efforts to date have often been specific to particular design flows and test cases. Another promising possibility is to carry out algebraic decompositions and mapping simultaneously, as in [LWG+97]. However, such an approach would still face the challenges of dealing with many possible placements for any decomposition, as well as exploring an even larger search space efficiently.

Thus, it is apparent that today's multilevel logic synthesis algorithms targeting routing congestion are inadequate and that there is scope for additional research in core multilevel logic synthesis algorithms targeting congestion as well as in finding good metrics that can guide such algorithms.

6.5 Final Remarks

In this chapter, we reviewed several technology mapping and logic synthesis optimizations that target congestion. Congestion-aware technology mapping techniques vary in the ease of implementation, runtime and memory overhead, as well as in the effectiveness. They range from simple augmentations to the traditional cost function used in mapping with a cost for the predicted netlength, to sophisticated congestion map based schemes that can guarantee delay optimality and minimize the area and delay penalty for congestion awareness by operating in congestion-aware mode only in regions that are likely to be congested. Although congestion-aware technology mapping algorithms have not yet made their way into industrial tools, they have yielded very promising results on a wide variety of industrial and academic designs.

The decomposition that precedes the mapping has a significant impact on the quality of the mapping solution. Although the work published so far in this area uses simple scalar congestion metrics, it is possible to extend decomposition to use congestion maps. However, it is not clear whether spatial metrics such as congestion maps can be made accurate enough at this stage to be able to provide consistent benefits. Other multilevel logic synthesis operations that precede technology decomposition include extraction, substitution, elimination, decomposition, and speed-up. In the congestion-aware counterparts of these operations, the objective function is modified by including some structural congestion metric. However, there is still significant scope for further research in this area. Given the powerful structural transformations that are available at this stage of the design flow, it is a promising direction for further exploration, especially since the routing congestion problem is likely to worsen with technology scaling and increasing design complexity.

References

[BHS90] Brayton, R. K., Hachtel, G. D., and Sangiovanni-Vincentelli, A. L., Multilevel logic synthesis, *Proceedings of the IEEE*, 78(2), pp. 264–300, Feb. 1990.

[CB04] Chatterjee, S., and Brayton, R. K., A new incremental placement algorithm and its application to congestion-aware divisor extraction, *Proceedings of the International Conference on Computer-Aided Design*, pp. 541–548, 2004.

[CP95] Chaudhary, K., and Pedram, M., Computing the area versus delay trade-off curves in technology mapping, *IEEE Transactions on Computer-Aided Design of Integrated Circuits and Systems* 14(12), pp. 1480–1489, Dec. 1995.

[DeM94] de Micheli, G., *Synthesis and Optimization of Digital Circuits*, New York, NY: McGraw-Hill Inc., 1994.

[GKS01] Gosti, W., Khatri, S. R., and Sangiovanni-Vincentelli, A. L., Addressing timing closure problem by integrating logic optimization and placement, *Proceedings of the International Conference on Computer-Aided Design*, pp. 224–231, 2001.

[GNB+98] Gosti, W., Narayan, A., Brayton, R. K., and Sangiovanni-Vincentelli, A. L., Wireplanning in logic synthesis, *Proceedings of the International Conference on Computer-Aided Design*, pp. 26–33, 1998.

[HWM03] Hu, B., Watanabe, Y., and Marek-Sadowska, M., Gain-based technology mapping for discrete-size cell libraries. *Proceedings of the Design Automation Conference*, pp. 574–579, 2003.

[KS04] Karandikar, S. K., and Sapatnekar, S. S., Logical effort based technology mapping, *Proceedings of the International Conference on Computer-Aided Design*, pp. 419–422, 2004.

[Keu87] Keutzer, K., DAGON: Technology binding and local optimization by DAG matching, *Proceedings of the Design Automation Conference*, pp. 341–347, 1987.

[KK03] Kravets, V., and Kudva, P., Understanding metrics in logic synthesis for routability enhancement, *Proceedings of the International Workshop on System-level Interconnect Prediction*, pp. 3–5, 2003.

[KK04] Kravets, V., and Kudva, P., Implicit enumeration of structural changes in circuit optimization. *Proceedings of the Design Automation Conference*, pp. 438–441, 2004.

[KSD03] Kudva, P., Sullivan, A., and Dougherty, W., Measurements for structural logic synthesis optimizations, *IEEE Transactions on Computer-Aided Design of Integrated Circuits and Systems* 22(6), pp. 665–674, June 2003.

[KBS98] Kukimoto, Y., Brayton, R. K., and Sawkar, P., Delay-optimal technology mapping by DAG covering, *Proceedings of the Design Automation Conference*, pp. 348–351, 1998.

[KS01] Kutzschebauch, T., and Stok, L., Congestion aware layout driven logic synthesis, *Proceedings of the International Conference on Computer-Aided Design*, pp. 216–223, 2001.

[KS02] Kutzschebauch, T., and Stok, L., Layout driven decomposition with congestion consideration, *Proceedings of the Design Automation and Test in Europe*, pp. 672–676, 2002.

[LWG+97] Lehman, E., Watanabe, Y., Grodstein, J., and Harkness, H., Logic decomposition during technology mapping, *IEEE Transactions on Computer-Aided Design of Integrated Circuits and Systems* 16(8), pp. 813–834, Aug. 1997.

[LJC03] Lin, J., Jagannathan, A., and Cong, J., Placement-driven technology mapping for LUT-based FPGAs, *Proceedings of the International Symposium on Field Programmable Gate Arrays*, pp. 121–126, 2003.

[LM04] Liu, Q., and Marek-Sadowska, M., Pre-layout wire length and congestion estimation, *Proceedings of the Design Automation Conference*, pp. 582–587, 2004.

[LM05] Liu, Q., and Marek-Sadowska, M., Wire length prediction-based technology mapping and fanout optimization, *Proceedings of the International Symposium on Physical Design*, pp. 145–151, 2005.

[PB91a] Pedram, M., and Bhat, N., Layout driven technology mapping, *Proceedings of the Design Automation Conference*, pp. 99–105, 1991.

[PB91b] Pedram, M., and Bhat, N., Layout driven logic restructuring/decomposition, *Proceedings of the International Conference on Computer-Aided Design*, pp. 134–137, 1991.

[PPS03] Pandini, D., Pileggi, L. T., and Strojwas, A. J., Global and local congestion optimization in technology mapping, *IEEE Transactions on Computer-Aided Design of Integrated Circuits and Systems* 22(4), pp. 498–505, April 2003.

[RV92] Rajski, J., and Vasudevamurthy, J., The testability-preserving concurrent decomposition and factorization of Boolean expressions, *IEEE Transactions on Computer-Aided Design of Integrated Circuits and Systems* 11(6), pp. 778–793, June 1992.

[SA93] Saucier, G., and Abouzeid, P., Lexicographical expressions of Boolean functions with application to multilevel synthesis, *IEEE Transactions on Computer-Aided Design of Integrated Circuits and Systems* 12(11), pp. 1642–1654, Nov. 1993.

[SSL+92] Sentovich, E. M., Singh, K., Lavagno, L., Moon, C., Murgai, R., Saldanha, A., Savoj, H., Stephan, P., Brayton, R., and Sangiovanni-Vincentelli, A., SIS: A system for sequential circuit synthesis, *Memorandum No.* UCB/ERL M92/41, University of California, Berkeley, CA, May 1992.

[SSS+05] Shelar, R., Sapatnekar, S., Saxena, P., and Wang, X., A predictive distributed congestion metric with application to technology mapping, *IEEE Transactions on Computer-Aided Design of Integrated Circuits and Systems* 24(5), pp. 696–710, May 2005.

[SSS06] Shelar, R., Saxena, P., and Sapatnekar, S., Technology mapping algorithm targeting routing congestion under delay constraints, *IEEE Transactions on Computer-Aided Design of Integrated Circuits and Systems* 25(4), pp. 625–636, April 2006.

[SIS99] Stok, L., Iyer, M. A., and Sullivan, A., Wavefront technology mapping, *Proceedings of the Design Automation and Test in Europe*, pp. 531–536, 1999.

[SSH99] Sutherland, I., Sproull, R., and Harris, C., *Logical Effort: Designing Fast CMOS Circuits*, San Francisco, CA: Morgan Kaufmann Publishers, 1999.

[VP95] Vaishnav, H., and Pedram, M., Minimizing the routing cost during logic extraction, *Proceedings of the Design Automation Conference*, pp. 70–75, 1995.

7

CONGESTION IMPLICATIONS OF HIGH LEVEL DESIGN

The traditional approach to congestion optimization consists of identifying routing congestion hot spots and then modifying the design locally so as to reduce the congestion in these hot spots. For this approach to be effective, the identification of congested areas in the design has to be reasonably accurate. While such accuracy is achievable at the late stages of the synthesis and layout, it is usually not an option early on during the design flow, when micro-architectural decisions are made and the design is floorplanned. At that time, few of the design blocks have been implemented. The only nets that exist in the design are a handful of global signal nets whose wire delays are deemed critical for accurate architectural performance simulations. Even the global clock and power distributions may not yet have been implemented. Therefore, any predictions of the amount of routing resources required in any particular region of the design are likely to be of little use for downstream congestion optimization. Indeed, while there have been numerous works on interconnect-aware floorplanning in recent years, there has been no successful attempt to date to model routing congestion during the floorplanning process.

However, many of the decisions made at these early stages in the design flow can have a significant impact on the eventual congestion in the design. The global wiring complexity of a design is heavily influenced by the micro-architectural choices made during early design exploration. These choices help determine the Rent's parameters for the wiring distribution of the design, thus determining the precise shape of the wiring distribution curve.

7.1 An Illustrative Example: Coarse-grained Parallelism

Let us illustrate the impact of the architectural choices for a design on its wiring complexity and congestion by considering the example of a hypothetical microprocessor being scaled to the next process technology node, using very simple first-order scaling assumptions. Let the wiring histogram of the processor at the current process node (drawn in solid lines in Figure 7.1(a)) be represented by the wiring histogram drawn in solid lines in Figure 7.1(b). Let us assume that the unit size along the (linear) horizontal axis remains unchanged across the two process nodes, but the unit size along the (logarithmic) vertical axis doubles at the scaled process node (so that a point (l, n) represents twice as many wires lying within the length bucket containing the wirelength l at the scaled node than at the original node).

Under classical scaling, in each technology generation, each of its linear dimensions will shrink by a factor of s (traditionally, equaling 0.7). As a consequence, the area required to implement logic equivalent to that existing at the current node will halve (assuming unchanged row utilization[1]). Since classical scaling does not change the shape of the wiring histogram of a design block (when normalized to the edge length of the block), the shrunk dimensions imply that the wiring histogram corresponding to the shrunk block (sketched in dashed lines in Figure 7.1(a)) will be represented by the dashed curve in Figure 7.1(b); observe that although the curve has shrunk along the horizontal axis, it is not compressed along the vertical axis (since the number of gates and nets is unchanged).

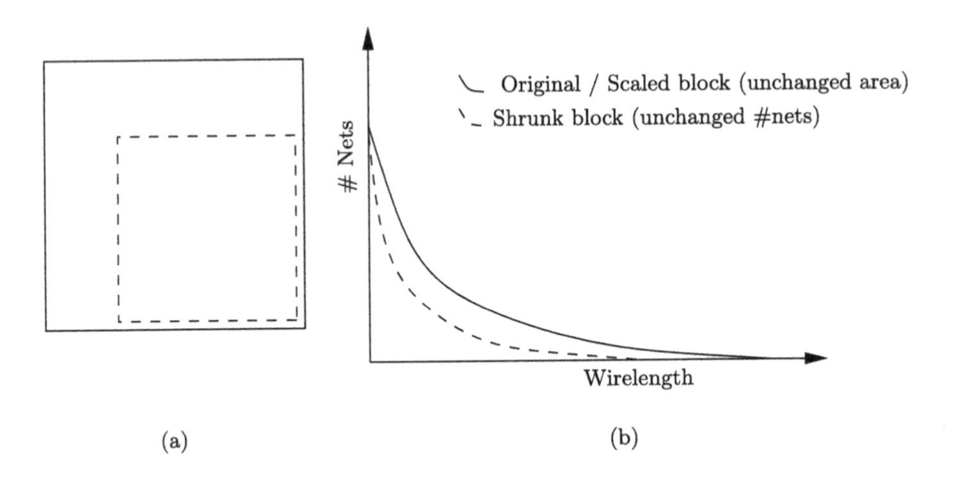

Fig. 7.1. Wiring histograms for the classical scaling of a generic processor across a process node.

However, as was discussed in Section 1.3.1 in Chapter 1, historical trends (as described by Moore's Law) show that the availability of smaller transis-

[1] In practice, the row utilization tends to reduce slightly with each process generation, primarily due to the increase in routing congestion with the growth in the number of cells within a block (as was discussed in Chapter 1).

tors and finer wiring architectures has invariably led to increased integration rather than to reduction in the area of the chip (or its non-cache logic). Therefore, it is reasonable to assume that the scaled processor will have the same area as its current generation, but will implement increased functionality by doubling its transistor count. In this scenario, the first-order scaling invariance of the wiring histogram shape of the block implies that this histogram is again represented by the solid lines in Figure 7.1(b) (even as it corresponds to twice as many wires as before).

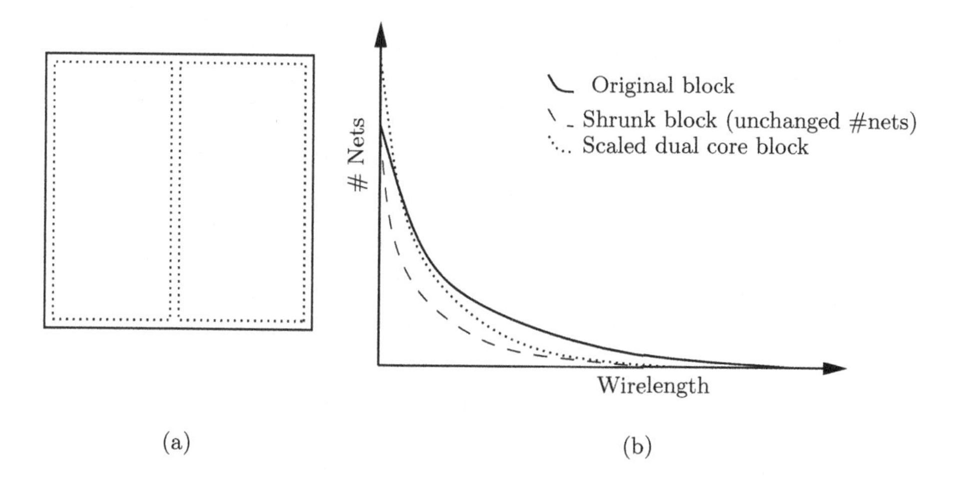

Fig. 7.2. Wiring histograms for the scaling of a generic processor to a dual core version across a process node.

An alternative to performance improvement through the functionality enhancement (*i.e.*. increased netlist complexity) implicit in the classical scaling picture is performance improvement through increased parallelism. Thus, instead of making the original core more sophisticated, it can be replaced by two loosely coupled cores each of which has the same complexity as the original core. While there is some throughput (and frequency) overhead for increased concurrency of this kind, it is often offset by the ease of implementation that enables a faster time to market for the scaled design. In terms of the wiring complexity, this approach corresponds to increasing the number of short, local interconnections while reducing the number of semi-global and global interconnections. This is illustrated in Figure 7.2(b), in which the dashed line represents the wiring histogram curve for a single scaled core, and the dotted line represents the wiring histogram for the entire dual core design. Observe that although the dual core alternative has approximately the same number of nets (and gates) as the classically scaled alternative (represented by the solid line in Figure 7.2(b)), the number of long wires (and consequently, the total wirelength) in the dual core version is significantly smaller; this translates to reduced average congestion in the dual core design.

Moreover, this reduction in the total wirelength of the design (and, in particular, in the number of long wires) has another significant benefit from the perspective of congestion. As was observed in Section 4.3 in Chapter 4, the length of a wire that can be driven by a typical buffer is shrinking much faster than the rate at which feature sizes are shrinking due to process scaling[2]. This leads to a rapid increase in the number and the fraction of buffers (and of nets that require buffering) in a design as it is scaled from one technology node to the next. We also observed that as the number of buffers in a design block increases, the buffers begin to have a serious impact on the routability of the nets within the block.

The number of buffers required for the wiring histogram depicted in Fig. 7.2(b) is significantly smaller than that for the wiring histogram for the scaled block shown in Fig. 7.1. This is primarily because the wiring distribution corresponding to the classically scaled design has a much larger number of long wires that must be buffered, as compared to the dual core design. As a result, the congestion caused by the buffers is also much more severe in the classically scaled design.

7.2 Local Implementation Choices

Gross architectural choices of the kind discussed in Section 7.1 impact the interconnection complexity by changing the *shape* of the wiring histogram of the design. At a somewhat finer-grained level, one can often reduce the routing complexity of a piece of logic by replacing the on-the-fly computation of a complex function in hardware by a cache that has been loaded with precomputed values of the function or by a lookup table implemented in hardware using a content-addressable memory; this technique works well for hard-to-compute functions that have a limited domain.

Indeed, most of the scheduling, binding and implementation choices made during the high level synthesis of a design impact the local congestion. Each of the alternative implementations of a functional unit results in a different local congestion profile. Thus, for instance, the decision to select a pipelined multiplier that requires three cycles instead of another one that can carry out the same computation in two cycles will change the local congestion in the physical neighborhood of the multiplier, since the interconnection complexities of these two choices are different. However, the cost vectors that drive such choices are usually limited to some combination of the predicted performance, power, and gate area.

[2] Under first-order scaling assumptions, optimal inter-buffer distances shrink at $0.586\times$ per generation, in contrast to the normal geometric shrink factor of $0.7\times$.

The routing complexity of these choices is sometimes known in isolation (when these functional units have already been implemented, or have been supplied as hard intellectual property (IP) by third-party vendors). Yet, it is not clear that their congestion impact when integrated within their parent block can be modeled accurately enough for the congestion to be usable as a meaningful cost during high level synthesis. Part of the problem arises from the thresholded nature of the congestion cost; a choice with a high interconnect complexity may be acceptable if its neighborhood is not very congested, but may not be routable if the wiring complexity in the neighborhood is higher than some threshold. But this wiring complexity of the neighborhood cannot be accurately determined until rather late in the design process, leading to a chicken-and-egg situation. In an attempt to ease this routability problem, some functional unit designers guarantee some routing porosity within their blocks up front. This increases the likelihood that their blocks can be successfully integrated without resulting in an unroutable design.

7.3 Final Remarks

In this chapter, we have seen that the architectural and micro-architectural decisions made during the early stages of a design can have a significant influence on the overall interconnection complexity (and therefore, the average routing congestion) of the design. Factoring in the expected cost of implementing the design (which includes the effort required for timing convergence as well as that required for routing convergence) at the architectural design stage itself can help avoid unpleasant downstream surprises. An architecture with reduced interconnect complexity is not only easier to route, but is also easier to converge timing on (because of smaller wire delays, fewer interconnect buffers required to compensate for resistive wires, and fewer cases of delay degradation due to unexpected route detours). Of course, an architecture with a low interconnect complexity can unfortunately still have local regions of severe routing congestion; however, the likelihood of such congestion hot spots is reduced.

The congestion is also influenced by the various choices made during the high level synthesis of the design. Yet, it is not easy to compute the local congestion cost of a selected functional unit accurately, even if it has already been fully implemented. This is because of the significant error involved in estimating the routing demands of the physical context within which the functional unit is to be integrated. Consequently, existing algorithms for high level synthesis do not try to optimize for congestion. Improved congestion prediction methods may make such optimization somewhat more feasible in coming years.

Index